AIRSTREAM

ZEN AND THE ART OF STREAMING

CHOOSING JOY

Publisher: Streaming Zen Labs, LLC
Author: Linda McCarthy
Executive Editor: Denise Weldon-Siviy
Interior design: Vanessa Moore
Artwork: Heather Wallwork
Cover design: Alan Clements

Airstream: Zen and the Art of Streaming

ISBN: 978-0-9817856-6-0

In loving memory
of my friend, Bob Shotwell.

CONTENTS

PREFACE

This book began with a gift.

When I sold my house in California, I had no idea where I wanted to live. A few years before that, a friend gifted me his Airstream. I began thinking (more seriously each day) about taking it on the road. That sparked a train of thought that's brought me countless miles, adventures, and joy.

Initially though, my gift brought more questions than answers. Could I really just live in my Airstream? Would it be enough to keep me safe? Content? Satisfied? Could I overcome my fear of being alone on the road? Would my friends think I was crazy? Would they be right?

A few days after I sold my house, I met a guy who helped me answer at least some of those questions. More accurately, I guess, he nudged me onto the path that brought me to those answers. Mr. Cool. Not his real name, of course. But a good description of his lifestyle. What isn't cool about Austin riverfront living? Funny enough, he owned several homes in multiple states. But for Austin, he wasn't ready to buy another new home.

It blew my mind that he could buy any house he wanted, but instead he chose to live in his Airstream. It was a choice we discussed often. And an opportunity he strongly encouraged me to take for myself.

He convinced me to dive into Airstream living and said, "It'll be one of the best things you've ever done." At the end of that pivotal conversation, I decided to get in the Airstream and GO. When a friend asked where I was moving to, I said, "To the land of abundance." He asked, "Where is that?" I said, "I'll let you know when I get there!"

Like so many of us, I'm no longer tethered to an office or frequent in-person work. It was time to take advantage of that freedom. I traded my cable/Internet bundle for a mobile hotspot and took off.

And WOW! After only two nights in the Airstream, I was sleeping peacefully. Deeply. That was a gift I'd given up hope for after a decade of fretful sleep.

It was also a gift that kept giving. Being rested kept me up for the adventure. When I hit the road in the summer of 2021, I met with friends I had not seen for years. I also met amazing new people all over the country. I saw more states than I had visited in the past decade. I broke through the fear that was keeping me from traveling alone. I found peace, joy, and adventure.

Along the journey, I also learned some very cool things about the Airstream. Important details about its aerodynamics, its protective design, and about the Airstream culture.

I hope some of the things I learned will help you. Things like how to best use your Airstream for staying safe, taking breaks from technology, protecting yourself from that same technology, reaping the benefits of being off the grid, being prepared for the journey, and so on.

That's why I wrote this book. If you are looking for the land of abundance, peace, joy and adventure, this book is for you. If you are thinking about getting an Airstream, this book is for you. If you already have an Airstream, this book is for you too. And like my friend did for me, I hope I can encourage you to get your Airstream out, really use it, and live your life.

ABOUT THE AUTHOR

Linda McCarthy was knocked down at the height of her career. While in Singapore on a book tour, Linda was poisoned by a fish dinner that turned out to be ten times more toxic than arsenic. The neurotoxins left her temporarily blind, debilitated to the point of being unable to climb a single flight of stairs. It took months of painstakingly brutal work to recover enough to function professionally. And more than a decade to recover completely. Along the way, Linda did a lot of hard thinking about what it was she wanted out of life.

While Linda was recovering, her best friend Bob Shotwell gave her an Airstream Bambi. At the time, she was much too weak to take the Airstream on the road. But through her recovery, just knowing that Bambi was there, waiting for her, kept her going.

By 2021, it was time. Linda sold her home in California and jumped into the Airstream life full-time. Her Bambi became Linda's home and office on wheels. Together they adventured from Florida to Louisiana, Texas, New Mexico, Colorado, Wyoming, Montana, Idaho, Utah, and then back to Texas. Along the way, Linda addressed her fear of traveling alone by learning as much as she could about Airstream security and applying those lessons to her travels.

Understanding security comes naturally to Linda. For almost three decades, she worked in the security industry, breaking into computers and testing security for Fortune500 companies. She took that background and its concerns with her on the road.

Staying safe and secure in your Airstream is just as important as protecting your computer, your corporate network, and your home. For full-timers, your Airstream rig *is* your home, even if it is on wheels. Planning ahead with security in mind makes that home a safe place for you to find peace and joy. As Linda's friend JP said, "Living in an Airstream is a lifestyle choice and being miserable is not an option." Planning for security allows you to follow JP's lead and choose joy.

Choosing joy is why Linda remains on the road and in the Airstream lifestyle. She explains, "Yeah, we do live in challenging times. I can't change that. But I can choose joy, so I do. Every day, I work to bring good energy into my life. Streaming helps me to do that in many ways. It's become part of who I am and the kind of calm peaceful life I want to have."

Bambi has since moved on to other pastures, but Linda continues to work and travel in her brand new 2022 Flying Cloud (named Bailey).

ACKNOWLEDGEMENTS

A special thanks to John Pigott who gave me an office to write this book in. He encouraged me to take my Airstream on the road and truly live life again. I doubt I could have taken this journey without his friendship and support. And if I had, it wouldn't have been nearly as much fun!

I'd also like to thank the professionals whose skills brought the mechanics of this book to life. First, thanks to my editor, Denise Weldon-Siviy. I've been fortunate to work with Denise for several decades. She's a great editor and a good friend; I just love working with her. More thanks to Vanessa Moore for her amazing-layout design and Heather Wallwork for her artwork. Heather's art always brings a book to life. And finally, thanks to Alan Clements who always comes up with brilliant cover designs. Thanks again to a great group of people. You all made this book possible.

1

JUMPING IN?

We've come a long way since Wally Byam, founder of Airstream, began selling plans for constructing his travel trailers for $5 apiece. Granted, that was during the 1920s, and Wally had designed his trailer to accompany his Model T. Even so, the thought of even a primitive DIY Airstream for under $100 is astounding.

By 1931, Airstream the company was born, bringing great joy to those of us filled with travel lust but who shudder at the word Ikea, envisioning newly drilled holes and hands full of spare screws and hardware. Still sporting the riveted aluminum body of the 1936 Clipper, Airstream continues to lead the pack in travel trailers, as it has for roughly a hundred years now. Operating continuously, excluding a short hiatus during World War II, Airstream was the only one of over 400 travel trailer companies to survive the Great Depression.

Fast forward to the early 2020s, and Airstream is perfectly positioned to meet the needs and dreams of a rapidly changing world.

Zoom Away?

Between 2020 and 2021, the way that Americans lived and worked changed dramatically. Almost overnight, we transitioned from a nine-to-five office economy to an expectation that the vast majority of professionals not only could work at home, but were surprisingly efficient at doing so.

Taking advantage of the Zoom work life, droves of young professionals abandoned the high rents and dense populations of urban centers. This actually continued a trend already in progress, just at a substantially higher rate. For the past five years or so, there has been a pretty steady stream of Americans leaving urban centers for greener (or at least a little less concrete) pastures. From 2017 to 2019, the net exit of people from American urban areas averaged 28,000 a month. In 2020, that doubled to 56,000. Those recently ex-city dwellers give many reasons for fleeing the city. Some express a strong desire for more green spaces. Others yearn for more space after spending far too much time inside. Still others cite their disappointment in urban living with so many amenities on hold due to health regulations.

Not surprisingly, when asked why they made the move, almost every single one fell back on the same simple sentiment—***Because I could!***

Consider the Road Less Traveled

While some of those urban escapees simply relocated to the suburbs, a surprising number opted for the open road instead. And why not? With travel a bit complicated recently, many of us are more than ready to set off for sights unseen. Or at least not seen recently.

Outdoor travel brings with it a myriad of benefits and advantages, many of which are both facilitated and complemented by the Airstream experience.

Forest bathing

Possibly one of the best things you can do for your health in the next year is to point your Airstream at the nearest forest, with forest bathing high on your mind.

"Shinrin-yoku", which translates roughly as "forest bathing," is the Japanese concept of soaking up the forest atmosphere—immersing yourself in the sights, sounds, smells, and experience of nature. Coined by the Japanese Ministry of Agriculture, Forestry and Fisheries in 1982, forest bathing is a concept that has seen astounding validation since its inception. It turns out that spending time in nature isn't just refreshing. It's incredibly healthy. Even short-term visits to urban nature centers have been scientifically proven to reduce stress. Spending quality time in nature has been shown to improve memory, lower blood pressure and heart rate, reduce anxiety, speed healing, improve immune response, and help to lessen depression.

Interestingly, the positive affects of forest bathing are mostly independent of the benefits of exercise in hiking or pursuing active outdoor sports. Simply spending time in nature, "soaking in" the forest ambiance, provides health benefits above and beyond the physical exertion.

Spending time with friends and family

Like forest bathing, spending time with friends and family can have wonderful benefits for your health. And the health of your entire family. Spending time with family alone has been linked to lowered stress, increased self-confidence, and lower incidence of mental health issues. And that's just for you. Your kids will benefit from better academic performance, lowered risk of behavioral problems, and the opportunity to learn appropriate conflict resolution. For all of you, having healthy relationships with either friends or family can increase your life expectancy by up to 50%. You read that right—FIFTY percent. That's confirmed by a meta-study of over 300,000 people reported in *Scientific American* which found that having a solid social network increased longevity by up to 50%.

Using your Airstream to connect the dots of that social network can be priceless. In the case of spending time with family, having an escape hatch can also be helpful. That's especially true if your family doesn't always model those appropriate conflict resolution skills. Airstream travel allows you to visit easily while still maintaining your own private space. Literally, the best of all worlds!

Finding peace, joy, and contentment

Study after study finds that happiness is more about being than having. Consumer toys, money, etc. are unlikely to bring lasting happiness. In the long run, what most of us are looking for is fairly intangible. Peace. Joy. Contentment. All of which are far more closely associated with experiences than things.

Experiences are what you will find along your Airstream journey. For part-time travelers, that could include forest bathing in all of your state parks over weekends. Perhaps spending a summer criss-crossing the country to visit old friends and long-lost relatives. If you're ready to embrace the Airstream lifestyle full-time, you could experience the joy of letting go of excess things and completely focusing on the journey.

Bringing Fido along for the ride

Another great advantage to the Airstream lifestyle is that pets are always welcome in your rig. We all know that pets, like forest bathing and social connections, have outstanding positive benefits to your health. Just the daily ritual of walking a dog greatly increases the odds that you'll meet the recommended level of physical activity for healthy living. On top of that, pet ownership reduces stress and anxiety, decreases loneliness and depression, and reduces the incidence of many types of mental illness.

Sadly, attempting to travel with pets generally does not lower stress or anxiety. Few motels allow pets. The few that do often have only a small number of pet-friendly rooms allotted. Quality can also sometimes be an issue. I've known friends to miss out on weekend travel opportunities because they couldn't find a pet-friendly accommodation or a kennel.

With Airstream travel, that's not an issue. You set the quality standards. And pet friendly accommodations are always available!

Discover the Airstream Lifestyle

Airstream isn't just a lifestyle. It's a culture. Like the Harley owners who congregate regularly around the country, Airstream owners have their own mobile community. And it's growing. The Facebook group *Airstreaming for Newbies* boasts over 20,000 members. Some full-time, some part-time. But all enjoying the ride.

Airstream culture

A good number of Airstream owners choose to connect through Airstream Club International. A member-only club, this group connects over 16,000 Airstream owners worldwide with service discounts, group activities, and each other. One of the member perks is access to over 900 free parking offers (called "courtesy parking sites") from fellow members for Airstream owners on the go. Even if you choose not to join the club, their website (*https://airstreamclub.org/*) is a great source of information about upcoming Airstream rallies and caravans.

Joining the Airstream culture, either in person via clubs or online to make connections, can be a first step in using your Airstream to meet new friends along your journey. Or you may choose to just show up at rallies and caravans and wing it. Concerned about being shy? Many other Airstream owners aren't. You can expect friendly welcomes and behavior most anyplace that Airstreams naturally gather.

Full-time or part-time?

In 2020, over half of factory new Airstream purchasers were new to the RV life. That number increased in 2021 and continues to increase in 2022, with backlogs of as much as a year between demand and production.

Surely, all of those newbies weren't ready for full-time RV living? Of course not. With an average income of $150,000, a surprising number of Airstream owners work while they travel (or perhaps, travel while they work). Yet others are retired or just determined to live off the grid.

The range of customers buying new Airstreams is nearly as varied as the Airstreams themselves. With 2022 prices ranging from just under $44,000 to well over $200,000 with options, it's a big range. New RVers may choose to begin small as weekend warriors and gently explore the Airstream lifestyle. Or they may take my approach, ditch the stuff, sell the house, and jump in full speed with both feet. Either way, they'll find themselves with plenty of companionship.

Glamping

For many of us, "camping" conjures images of drenching rain, pernicious insects, and—let's be blunt—deprivation. Fading to sleep to the pitter patter of gentle rain only to awaken on rocky ground in a growing puddle as another air mattress bites the dust and sleeting rain pours right through the tarp carefully laid under your tent. For this reason, some of your friends simply won't understand your Airstream investment. To truly explain, you'll need them to understand that camping is not even close to "glamping" (which is derived from *gl*amourous + *camping*).

Glamping is the boujee upscale version of camping. It's camping without the muss and definitely without the deprivation. You might be parked at the same campground as those poor tent campers, but inside your Airstream, you're living in a different reality. One with air conditioning. A private bath. A 60-inch HD projection TV. Adjustable beds with high-end memory foam mattresses. Blackout curtains in the bedroom area. Even high-tech controls that let you extend or retract the awning and adjust lights from your smart phone. At 30 to 33 feet, the 2022 Classic Travel Trailer, designed for full-time living, provides separate cooking, dining, lounging, bathroom, and bedroom areas. High-end models come standard with Egyptian cotton sheets and a cedar-lined wardrobe. This is definitely not your father's camping.

Make Your Choice. And Name It!

When you're ready to dive in and choose your adventure, keep in mind that you get to name it as well. Officially, that is. Airstream maintains an official registry. When you register the name of your Airstream, you can order an official Medallion inscribed with that name and join the virtual caravan of registered Airstreams.

Before you take the naming plunge, put some thought into it. Sure, *Silver Bullet* and *Tin Can* were original for maybe the first generation of Airstreams. *Fullmetal Chariot* and *Rosie the Riveted* are a bit less overdone, but still showcase someone else's style and sense of humor. You'll probably want something a bit more personal. Something that speaks of *your* plans for the Airstream lifestyle. Take your time.

Also consider yourself lucky at only needing to name a single Airstream! Consider the case of Elon Musk. While we suspect he delegated the honor, Musk has had ample opportunity for Airstream naming. Back in 2016, Tesla rolled out a small fleet of Tesla Model

X SUVs, pulling Airstreams configured as mobile design studios for consumers to order their own Model Xs after checking one out in person. He later increased his Airstream footprint by adding an RV park's worth of Airstreams as housing for employees at his Space-X launch base in Boca Chica, Texas. The man himself? No, he doesn't live in an Airstream there. But he does live in a 375-square-foot tiny house. In comparison, a 33-foot Airstream, with over 100 square feet more than he's accustomed to, would feel downright spacious.

Ask Yourself

- Do you see an Airstream in your future? Your NEAR future?
- Is it time to try part-time glamping?
- Are you ready for mini-adventures? Are you craving adventure?
- Do you have a list of friends or relatives you've been "meaning to visit sometime"? Is that sometime now?
- Are you seeking to get away from things and routines?
- Are you ready to consider life on the road?

If you answered yes to any of the above, then it may just be time for jumping in!

2

DON'T BE A TARGET

Finally, you have your own rig! Keeping that rig safe is now your responsibility. With crime on the rise in 2022, keeping yourself, your family, your pet, and your rig safe must be a priority. In this chapter, I cover some of the things you can do to keep your rig safe both on and off the road. In Chapter 3, I cover how to keep yourself safe as well. In both instances, staying safe takes planning.

In many ways, keeping your rig safe is much like protecting your home. If you've chosen to adopt the Airstream lifestyle full-time, protecting your rig IS protecting your house. The big difference, of course, is that a criminal can't drive away with a traditional house. For brick-and-mortar homes, your home security focuses more on protecting the belongings in your home than protecting the home itself from theft. With your Airstream, you need to think about both.

No security system is 100% foolproof. However, using layers of security can protect your rig from break-ins and keep it from becoming a target for RV thieves.

Recognize the Risk

Nothing can ruin your Airstream experience faster than a break-in or even worse, the theft of your RV. And those thefts are definitely on the rise. That's hardly surprising given the huge jump in demand for RVs.

Demand for RVs under Covid has been especially high with over 430,000 sold just in 2020. While final figures are always lagging a few months, industry forecaster RVRoadsigns estimated total 2021 year-end deliveries around 602,000 units—almost 40% higher than 2020. Demand is expected to keep increasing by double-digits throughout 2022. No wonder criminals are looking to target this growing opportunity.

Since RV security break-ins and thefts are lumped into other statistics, we need to look at the overall crime rate to get an idea of the severity of the problem. Annually, on average, 2.5 million house break-ins happen in the United States and surprisingly most break-ins occur during the day. And RV thefts are more common than you would think. As early as 2017, long before the pandemic spurred historically high RV purchases with up to year-long waits for delivery, RV thefts already accounted for 6% of stolen vehicles and were the fourth most common type of vehicle stolen. While RV thefts are still very uncommon, the number of reported thefts jumped 867% in 2020. Given the high value of RVs, that's hardly surprising. In March 2020, thieves at Skaggs RV Outlet in Elizabethtown, Kentucky made off with two RVs valued at over $70,000 in only 10 minutes.

Layer Your Security

Buying a new rig is like buying a new home. You need to check all entry points in your home and harden your defenses. You might add an alarm system or install your own security cameras. Or both. Are you in a high-crime area? If so, you need more protection than in a lower-crime neighborhood, but these are all things to consider with every new home purchase.

I've spent the last 30 years working in computer security, advising Fortune 500 companies how to protect themselves from theft and intruders. Not surprisingly, when I got my new Airstream, I immediately began wondering about security. What were the risks? And more importantly, how could I reduce those risks?

One of my first realizations was that securing my Airstream wasn't all that different from securing computer networks. A bit heavier on the physical protection side, but the process was similar. I began by making lists of every way I could think of that someone could make off with my new rig or its contents. Then I began investigating products to shut down each of those potential routes. What I found was similar to computer security—that I would need several layers of security. Each layer alone would provide some protection against theft. Some minor. Some major. But *together*, these layers dramatically decrease risk.

You may choose to add all these layers or only some. That will depend on your comfort with risk, and the logistics of where you plan to travel or park your rig.

The top security layers to consider are:

- Outdoor lighting with sensors
- Window and door alarms
- Indoor and outdoor cameras
- Receiver hitch lock and coupler lock
- Wheel lock (boot)
- Safe for valuables
- Locked outside compartments
- Steering wheel lock
- GPS tracker
- Smoke and carbon monoxide detectors
- Glass break detector
- Motion detector

Outdoor lighting with sensors

Even though many RV thefts happen during the day, they can also happen at night. The best way to deter night-time thefts is with outdoor lighting. If I'm staying in an RV park for an extended period, I have string lights that I hang outside and keep on a timer. They come on every night and help keep my space well lit. I also have motion detector lights that will come on if someone comes near my rig. It's part of my overnight safety precautions to keep lights on anyone who happens to wander into my space.

It's important to keep your space well-lit while you're traveling. Simple string lights are easy to hang when you're staying at a location for more than a few days. I sometimes even string these lights up into the trees in my space by taking a water bottle and tying it to one end of the string. Simply toss the water bottle into the tree, pull it down, then toss up into the next tree. It looks great at night and people stop by and ask how you got the lights up that high in the tree. For a bit of fun, you can use a string of twinkle lights, giving a festive veneer to your campsite, as if you were preparing for a wedding or forest birthday celebration.

Keeping your space well-lit makes your space enjoyable and helps to keep any creeps from lurking around your rig. For short stays, especially overnight, you may not be up for throwing twinkle lights into the trees. An easy shortcut is to just set out a few inexpensive solar LED motion sensor lights. I found great battery operated lights for under $25 on Wayfair.com. When you're looking, check the reviews closely—not just the price. While I settled on the $25 lights, I'd first tried out much more expensive lights that didn't work nearly as well.

Window and door alarms

Window and door alarms provide an additional layer of security and come in a wide range of levels. Some are simple, generating a loud audible alarm tripped by sensors that detect door tampering and/or glass breakage. Some systems tie into Smart phone apps. Others are standalone. Some tie into video monitoring systems. Some don't. The range is nearly as wide as the range of available Airstream options.

For details, talk with your Airstream dealer and check out the AirForums online (*https://www.airforums.com/*) to see the range of solutions that other Airstream owners have come up with. One lower-tech solution you'll see referenced multiple times is just a reminder that Airstreams are very pet friendly. A large dog nearly always trumps a loud alarm. Less volume, but more teeth!

If Fido isn't an option, consider that alarms are very effective at reducing risk. In one University of North Carolina study of incarcerated burglars, 60% of convicted thieves said that if they found an alarm system or camera, they'd move on and select a different target.

Indoor and outdoor cameras

Camera surveillance, *especially* if it's obvious, is a great deterrent to theft. And to be honest, the more obvious, the better. TV shows routinely show thieves disconnecting camera systems. In real life, however, thieves more often move on to a home or vehicle without a security system.

When considering the type of security system to install, keep in mind the real cost and not just the initial outlay. In addition to the reduced risk of theft, installing a security system might save you up to 20% on your insurance premium. Also keep in mind that, like the other security measures, cameras also work best in combination with additional security measures such as increased lighting and coupler locks.

Some security cameras connect using WiFi or a cellular network like Verizon, T-Mobile, or AT&T. The options you have for these security cameras are limited to those supported by your cellular provider. Security cameras that use Verizon often won't work with T-Mobile and vice versa. Because of that limitation, you'll want to start by investigating which security cameras are supported on your cell plan.

The security cameras I choose, might not be what you need. Here are a few different types of cameras to consider.

Vosker Wireless

The Vosker V200 mobile security camera is 100% wireless and uses 4G/LTE mobile connectivity (no WiFi required). It has a built-in solar panel and requires batteries for backup power. This camera is a good solution for anyone living off-grid or parking in remote locations. Videos are recorded locally on an SD card and photos are sent to the mobile app.

This camera is pricy but if you are living in a remote location, it could be the right camera for you. It has a free starter plan with 100 photos per month, or you can test out their 30-day trial plan. Unless you're at a construction site with a lot of people coming through (which is unlikely to be remote enough to be off-grid), 100 photos per month should be more than enough. This is a great camera to monitor your outdoor space. If someone enters your space, the camera will take a picture and the cellphone app will alert you immediately. It takes super clear pictures both day and night.

Blink Security

Blink WiFi cameras are versatile and inexpensive. They offer indoor and outdoor cameras with both day and infrared vision. Both types are wireless and come with two lithium batteries that last up to two years. The outdoor cameras can be installed with a solar panel to help keep them charged. The Blink Home monitoring app allows you to hear and speak in real time. You can customize zones to monitor and receive motion detection alerts on your cellphone.

With a subscription plan, you can store your video clips and photos in the cloud. Or you can store video clips and photos locally on a USB flash drive with the Blink Sync Module.

In addition to the Blink cameras, Blink also offers a wireless, battery powered floodlight. The Blink Floodlight is motioned-triggered with 700-lumen LED lights. The monitoring app will also send notifications to your cellphone when the floodlight is triggered.

Both the floodlight and Blink cameras provide good coverage. Blink security is definitely DIY. If you have problems installing the cameras, you might need to solve it on your own. I attempted to call their support line, but nobody answered. Ever. If you're comfortable with technology, the lack of support might be a fair tradeoff for a security system that can be set up for only a few hundred dollars.

Receiver hitch lock and coupler lock

The receiver hitch lock and coupler lock serve two purposes. The coupler lock prevents a thief from hooking up your parked Airstream to their vehicle and driving away with it. The receiver hitch lock prevents a thief from unhitching your attached Airstream (at say, a gas station, or parking lot) then hitching it up to their vehicle and driving off.

In many ways, these locks are like bicycle locks. You want to make sure you purchase one that would be difficult to hack or saw off. I tried out several. Some were easier to saw off than others. Some had locks that were easy to pick. Some didn't. One more expensive device, Proven Locks 2516, took three days for a lock picking expert to hack. This is not an area where you should cut corners. It's a one-time cost that can make it difficult for a criminal to smash off with a sledgehammer, saw off, or pick the lock. Like thieves, security solutions are always in motion. I keep testing out equipment so I might be testing a different coupler lock next week. However, I never go without one.

You'll notice that I said "difficult" to saw off, not impossible. No security product is ever 100%. But some give you far more time than others. My Dutch friends, who are quite the experts on bicycle locks, refer to specific styles as 30-second locks versus 15-minute locks. While a thief is trying to make off with your trailer in a public parking lot, that extra time is a serious deterrent. Anything that makes it harder for an RV thief to take your rig is worth doing.

Wheel lock (boot)

When I'm traveling to a new location or away from my rig for an extended period, I add a wheel lock. Sometimes called a boot, the wheel lock is similar to the bright yellow boots that police use to disable cars in no-parking zones, or vehicles with massive numbers of outstanding parking tickets.

While wheel locks come in various colors, I really like the neon yellow. It's like a poison dart frog whose neon colors tell potential predators that this frog is protected. My yellow wheel boot is a bright visible sign that says, "This trailer isn't going anywhere easily!"

Like the coupler lock, the wheel lock isn't foolproof on its own. It's just one more layer of security that makes it harder for someone to drive away with your rig. If you're adding all the security options I've mentioned so far, at this point a thief would have to stand under your attention-grabbing lights but out of view of your surveillance cameras as they try to saw through your coupler lock and then go on to remove the wheel. Dealing with each new layer of security takes more time and more energy. Many thefts are opportunistic. The idea is that at some point the potential thief will realize that stealing your trailer is just *too much work* and he'll move on down the road to look for an *easier target*.

Safe for valuables

While protecting your rig is essential, you also want to keep the belongings inside your rig safe. Your first line of defense there is simple—keep strange eyes from peering into your space. That means not inviting strangers inside and keeping the curtains closed when you're not in.

Even with your curtains closed, you should keep your valuables hidden. Take it up a notch and bolt in an actual safe. If you have a new Airstream, you should work with an approved repair center to have them install a safe correctly so that you don't void your warranty.

Locked outside compartments

This isn't so much a security feature as a security procedure. Every Airstream comes with locks on the outside compartments. Not every Airstream comes with an owner who remembers to always keep those locked. I really had to train myself to do this. You should too. Be that security conscious owner!

Steering wheel lock

A steering wheel lock is one more level of deterrent. Easily installed and locked, this device freezes the steering wheel making it near impossible for a car thief to steer your car if he's managed to get past your other security measures and into the car itself. This will prevent a thief from driving off with both your rig and the vehicle you use to tow it. Like the other security measures, it works best in conjunction with other layers of security. Keep your tow vehicle locked at all times.

GPS tracker

If all else fails and a crazy determined thief makes it through all the above-mentioned security features, a GPS tracker will capture their every move. Or more literally, it'll tell you—and you can tell the authorities—exactly where the thief has taken your rig.

Numerous GPS trackers are available. I like having more than one tracker. I have an inline GPS tracker that a criminal won't know is installed. I also like having a removable tracker. This way, if a criminal gets in and finds one tracker, they won't find the other. You can purchase a battery powered device that attaches magnetically in a hidden location. A magnetic strip on the device will allow you to hide it somewhere that makes it difficult for someone to find if they break in. I've tested some that have long-lasting rechargeable batteries that last up to six weeks and will notify you when the battery needs charging.

Built-in alerts like these will let you know if the device is tampered with or removed. A good GPS tracker will allow you to configure a geofence around your Airstream and will notify you if the device leaves a predefined area such as where your rig is parked. While your rig is parked, the GPS tracker should wake up occasionally to check its position. Once your rig is in motion, however, the device will stay awake, continuing to track location by GPS and posting that information to a smart phone app.

Smoke and carbon monoxide detectors

Theft isn't the only bad thing that could happen to your rig. While we keep talking about your trailer as if it were a vehicle, it's really a house on wheels. That's a house with a kitchen and heating system. That means that house fires are entirely possible. Because Airstreams carry propane, which is highly explosive, trailer fires can be especially dangerous. Even a small 3-pound propane container could explode with sufficient force to destroy a trailer. Not to mention the damage it could do to surrounding RVs if parked in a campground at the time.

For this reason, new Airstreams come with smoke and carbon monoxide detectors and a fire extinguisher. You need to make sure those items are accessible and kept current with fresh batteries. At delivery, the smoke detector will not be activated. You'll need to install the provided battery and test.

Designing Your Security System

It's easy to set up a brick-and-mortar home security system because houses don't move from one place to another. RVs need security systems that can move and work at the next location. Security systems that rely on Internet or cellular connectivity can also be challenging depending on where you plan to stay. In some locations, you may not have cell service or Internet access.

The security system I choose may not be the right security system for you. Before you can research to find the right security system for your Airstream, you need to decide which features are most important to you.

Some questions to consider are:

- Does the security system require WiFi?
- Does the security system connect via cellular network?
- Will your RV be parked at one spot or frequently moving to new locations?
- Will you need a hotspot or device to connect to a cellular or WiFi network?
- Do you want cellphone notifications of security issues?
- How difficult will it be to set up?
- Are cameras all you need?
- Can you get customer-service support if you have questions?
- How much does it cost?
- Is this a one-time cost or are you paying cloud storage fees?
- Do you want cloud storage?
- Are door and window alarms important to you?
- Do you want to monitor for glass breakage?
- Do the security cameras have night vision?
- Do you want a monitoring service?

- Do you need to monitor the temperature inside your RV? If you camp with a pet, this could be useful to make sure that the RV doesn't become too hot or cold for your pet's safety.

Some additional things to consider are connectivity and monitoring. Before selecting a security system or individual security devices, you need to know how they connect. For example, the Wingard Connect 2.0 device connects to WiFi signals or to a 4G LTE data plan, allowing you to connect all your devices. (They also plan a 5G version but haven't yet announced the release date.) Not all devices offer both WiFi and data plan connection. Likewise, the Wingard device supports multiple carriers such as AT&T, Verizon, and T-Mobile. Check with your carrier to make sure this device is supported and do your own research carefully before buying.

In addition to considering connectivity, you'll need to decide whether you want active monitoring and support. When you are on the go, a monitoring service might not make sense. If you are parked and living in your rig, you might want a 24/7 monitoring service. Also, if you are visiting an area for six months or longer, you might want to turn on the 24/7 monitoring service for extra safety.

One option for a robust monitoring system is Simply Safe. Like Blink, Simply Safe is DIY, allowing you to customize based on your needs. You can include a wide variety of devices, including indoor cameras, wireless outdoor cameras, door entry sensors, window sensors, motion sensors, glass-break sensors, carbon monoxide detectors, smoke detectors, and temperature sensors. Some of these devices may not make sense for you. For example, if your rig doesn't have skylights, you may not need a glass-break sensor in addition to window alarms. The devices are inexpensive, and you can configure a good system for under $500.

With Simply Safe, you can pay for monitoring or use just the alarm. The alarm system includes a loud siren with a volume control that can be set to LOUD that will most likely scare off criminals. In the case of a very remote location, there may not be anyone around to hear a loud alarm, rendering it ineffective. Monitoring services are month-to-month so you don't have to commit to a contract up front. You can even test out their professional-monitoring service for free for a month and see if it's something you want to stick with.

Whatever security system you choose, remember that available security devices change frequently. This is an area you may want to return to and consider upgrading in the future. Already, there are far more security cameras on the market today than there were a few years ago. I expect RV security to continue to improve, especially in regard to mobility. As that technology evolves, I will continue to test and review new systems and devices. To keep up with my research, be sure to bookmark my website, *www.airstreamsecurity.com* (this site will be coming online soon).

Don't Void Your Warranty

If you purchased a new Airstream, make sure you don't install something that voids your warranty. For example, if you change your locks out and it's not approved by an official Airstream repairs center, you can void your warranty.

In general, you're safe selecting security devices that aren't hardwired into your Airstream. Most of the security features we've discussed fall into that category. The obvious external devices include motion sensor lights, steering wheel locks, wheel locks (boots), and magnetic GPS trackers. What's important in terms of the warranty is whether the added security device is hard-wired into your Airstream. If you're not sure, always check with your Airstream service department.

Factor in Location

Selecting the right camping site can also reduce the risk of your rig being stolen. While it seems counterintuitive, Airstream of Utah recommends on their blog that you avoid camping right along a main road. While that may provide a better lit area, it's also a lot more convenient for thieves. Their recommendation? The farther a thief has to travel to get to your RV, the less likely they are to do so.

You also need to protect your rig when you're not on the road. While you may personally feel more vulnerable when you're out and about, your rig is most likely to be stolen at a place other than a campground. Your rig is most vulnerable parked at a residence, restaurant—even an unattended mall parking lot. RVs are especially vulnerable while in storage. That's something to consider if you take the snowbird approach and travel summers while storing your rig in between adventures. Remember that you need to have the same level of vigilance at mundane places as you do in the wilds. The suburbs can be much more dangerous than you think!

Remember the Human Factor

All the security devices in the world won't keep your rig secure if you don't use them properly and consistently. A steering wheel lock is only effective if it's locked to your steering wheel while you're out. The same lock tucked under the seat and forgotten is a great idea that just wasn't implemented. If you're going to buy wheel locks, you need to make sure that you use them. For best effectiveness, that should be every time that you're away from your rig. And especially in high-risk areas like service stops. It's easy to think, "This seems pretty safe, I'll follow through on the security stuff later."

Getting past that human tendency to underestimate risk and procrastinate is the first half of what I like to think of as "the human factor." The second half is quite simply being aware of other humans.

Airstream owners frequently consider themselves part of a select, often tight-knit community. The many Airstream clubs, rallies, and caravans are testimony to that culture. It's a friendly group, filled with members who often share similar interests. One of the great jobs of the Airstream culture is connecting with the other members of the club.

Even when those connections are just in passing, they can also form an important layer of your security. Consider my friend Steve's experience. Steve had a full-time space in a well-known Texas RV park. While he was at work, someone pulled up to his rig and was getting ready to hook it up and drive it away. Luckily, Steve had two levels of protection in place that day. First, he had a coupler lock so the thief had to saw off the coupler lock before he could hook up his rig. Removing the coupler lock took time, time during which the would-be thief was in view of the neighbors. That was his second layer of defense. Steve had taken the time to get to know a friendly couple nearby who were living full-time in their rig. When the neighbors spotted someone preparing to hook up Steve's rig, they called him to ask if he was loaning out his Airstream to a friend. "HECK NO!" Luckily, Steve worked nearby and was able to quickly drive to the RV park and block the would-be thief in with his Land Cruiser until the police arrived. Sound brazen? Daytime thieves often are. This one told the police he had an order to tow the rig. When he failed to produce any proof of that, the Texas police cuffed the creep and took him away.

Steve's experience is a good example of the need to layer your security. The hitch lock alone slowed down the thief but didn't stop him. But it did drag out the process, giving the neighbors time to notice the attempted theft. It was the combination of those layers that protected Steve from losing a truly awesome Airstream with an important history to him—something that really couldn't be replaced.

Stranger danger?

Hmmm… You're probably wondering now, isn't part of my security not trusting strangers? And it is—within reason. It's possible to make new friends while safeguarding your Airstream in the same way you'd befriend neighbors while safeguarding a brick-and-mortar home. The tricks? Move slowly. Trust your gut. And be okay with protecting your private space.

When I first hit the road, friends who'd already adopted the Airstream lifestyle had warned me that people would often ask to come inside to see my rig. To be honest, I thought that was a bit creepy. I couldn't imagine a stranger walking up to my 4-bedroom brick-and-mortar home and asking to come inside to check it out. Sure enough, though, this happened to me the very first time I took my Airstream out. Two very nice seeming women asked if they could come in and look around. I gave them a polite "No", using the excuse that I was still organizing things and it was a bit of a mess. They didn't seem like a threat. I decided before I even hit the road that I wasn't willing to welcome strangers inside my Airstream. I realize that you might like inviting strangers in your rig. It's just not something I'm comfortable with.

Or new friends?

Luckily, keeping the socializing outdoors in no way really hampers my ability to meet great new people and make friends. Through my Airstream, I've met a ton of amazing new people. One of the first new friends was an older couple just a few spaces away on my second trip out. Learning that I was new to the Airstream life, they invited me up to their campsite and at the same time taught me an easy way to have new friends "over" while still protecting the privacy of my inside space. "No need to bring a chair," they said. "We keep extra chairs outside for people to stop by and sit with us."

I did stop up from time to time. In fact, I came to trust the new neighbors enough to let them know when I was going out of town and to ask them to keep an eye on things. I gave them my phone number in case they saw something strange. Like Steve, I now had friendly neighbors who'd let me know if someone randomly tried to drive off with my rig.

Making friends cautiously with the right people is fun and it helps form a personal safety net. Most RVers look out for their neighbors because they would want you to look out for them too. I meet so many nice people—they are out there—but they don't have to come inside. Remember, your rig is your safe zone, and you want to keep it that way.

Ask Yourself

- Am I aware of my surroundings and possible risks?
- Can I identify when my rig is most vulnerable to theft?
- Have I identified the security layers most important to me?
- Do I actually *use* my security features once installed?
- Am I combining enough layers of security to provide serious protection?
- Can I gracefully say "No" to requests that encroach on my private space?
- Am I open to cautiously making new friends?

3

PROTECTING YOURSELF

One of the great benefits of owning an Airstream is its safety record. The aerodynamic design provides smooth handling—a critical feature for the larger trailers. That same design, coupled with the trademark aluminum construction, may also provide you with physical protection from excess electromagnetic fields. Combined, these features mean that while you're taking simple measures to secure your rig, that rig is already busy securing you! In addition to the protections provided by the Airstream design itself, you can add to your personal protection with solid route planning and weather awareness, and by selecting camp sites with personal safety in mind.

Aerodynamic Design

Airstream is hands down the most recognizable brand in the recreational vehicle (RV) industry. Mention Airstream to most people, and they'll immediately know exactly which campers and trailers you mean. Even people with no knowledge or experience in the RV life recognize the silver bullet-shaped trailer. And no wonder. This iconic design has been around for nine decades.

Airstreams were truly designed to glide through the air. That focus on aerodynamics began with the very first Airstream that Wally Byam designed shortly after graduating college. The first Airstream that featured the "silver bullet" design hit the market in 1936.

Aluminum was lightweight and inexpensive in the '30s and Wally's shiny design took off. Production at the Airstream factory in California was steady with the lone exception of the World War II years when Wally closed the factory and went to work in the aviation industry. In addition to benefiting the war effort, the experience he gained in aviation had a profound impact on Airstream design as Wally modified the shape of the Airstream based on early aircraft design. To improve the Airstream aerodynamics, Wally rounded corners and reduced the weight. Not only did performance improve, so did gas mileage.

The engineering focus has continued through the years. In the 1970s, an aerodynamic study showed that Airstream was 20% more fuel efficient than typical square-shaped trailers. Fifty years later in

2022, Airstream researchers continue pushing the boundaries of Wally's passion for aerodynamics. The focus now is not only on gas milage but towing with an electric vehicle that drives Airstream's engineering team into a new decade of towing expectations. That means trying to take an already great design and pushing its boundaries to make it more economical to tow and to reduce its carbon footprint.

That continually refined engineering focus coupled with the futuristic design is what has placed Airstream ahead of the competition since 1936. Fast forward almost 90 years and today's Airstreams *still* have that modern, timeless, and futuristic look.

The Airstream look isn't the only thing that's durable. Seventy percent of all Airstreams ever built are still on the road today. No other vehicle manufacturer can make that claim. Although I had to wait six months for my custom Airstream, I knew that it would be hand crafted in America by a company that wasn't trying to speed up the assembly line. While the typical "white box" RV trailer takes 50 to 60 hours to produce, Airstream rigs require on average 350 skilled man hours as each detail, including the 3,000 rivets, is meticulously executed.

Electromagnetic Fields

I started working in computer security in the late 1980s, before most people even knew that the Internet existed. I viewed technology like Tim the Tool Man (another '80s icon) viewed power. The more the better!

I'm a geek at heart, so I still feel that way today. I've also started to question whether too much of a good thing is still good. And too much doesn't even begin to describe the proliferation of technology.

Today, 15 billion people have cellphones. For most of us, that mobile device includes smart phone technology and is rarely out of fingertip reach. Look around the next time you're out in public. How many of the people around you are playing on their phones? Checking email? Browsing social media? Engaging with any one of a million apps?

Cellphones are also just the tip of the iceberg when it comes to the Internet of Things (IoT). IoTs are all the Internet-connected devices you see so often that you've probably forgotten their connectivity. Your fitness tracker. The bathroom scale that automatically sends your weight to a fitness app on your phone. Baby monitors. Your home security system. The GPS in your car. Google home. The smart speaker. Air pods. Home appliances are increasingly including connectivity as a bonus function. The washing machine that texts you when the load is done and ready for the dryer. The camera-equipped smart fridge that allows you to look in the fridge from your office to help plan that market stop on the way home.

A surprising number of medical devices are also part of the IoT. If you're diabetic, your blood glucose monitor most likely transmits data to your health care provider. Pacemakers, insulin pumps, and other medical devices also transmit data directly, eliminating human error and providing more complete and more accurate health information. Some doctors' stethoscopes are also connected devices.

Why is this proliferation of IoTs concerning? Well, a couple of reasons, honestly. Some people worry that constant connection has adverse effects on mental health. Others worry about privacy. There's also a growing concern about possible physical damage caused by constant exposure to electromagnetic fields.

Electromagnetic field (EMF) is a term that includes the radio waves, electric fields, and magnetic fields that comprise cellular and other communications networks. It also includes "dirty electricity" which is EMFs caused by small (micro) surges that leak from power lines, household wiring, and electrical devices. Every day we are exposed to EMFs from all the electric appliances around us in addition to phones, computers, TVs, etc.

EMFs consist of photons that travel at the speed of light in a wave-like pattern. The energy found in photons form different types of electromagnetic radiation and are expressed in energy, wavelength, or frequency. Frequency is measured in hertz (cycles per second), wavelength measured in meters, and energy measured in electron volts. This is why you'll sometimes see concerns about EMFs referring to electromagnetic "radiation." You'll also see references to radio frequency waves (RF). RF is a form of electromagnetic radiation which is a combination of electric and magnetic fields that transfer energy in the form of radio waves.

Because smart technology is now everywhere, many people are starting to question its safety. Historically, we've seen that EMFs are safe in small doses. But are we still getting "small" doses given the huge numbers of connected devices? It's a valid question. After all, the difference between medicine and poison is often dosage.

What's the risk?

Obviously, EMFs aren't fatal at the doses we see today. Smart homes aren't imploding. Remote tech workers aren't dropping like flies into their kombucha and avocado toast. But that doesn't mean that EMFs are harmless. Multiple studies are ongoing trying to identify the level of risk.

Initially, studies asking about health and technology focused on the EMFs generated by cellphones. An entire generation of moms used to warn kids not to sit too close to the TV. Cellphones, of course, changed that dynamic. We weren't sitting closer to the technology. Some of us (teens in particular) spent half the day with that technology pressed up against their check. Exposure was much closer and of longer duration. This led to speculation that cellphone use might increase brain or neck tumors, prompting numerous studies of that very issue. National Toxicology Program researchers ran several studies looking at tumor incidence in rodents exposed to cellphone radiation. The results? Mixed. In 2016, researchers sounded the alarm at the tumor link. By 2018, the consensus was that little to no link existed. By 2020, a 2-year follow up was decidedly indecisive. That latest study found a clear link between cellphone radiation and heart tumors in male rats. The study also identified a possible link for brain tumors and adrenal tumors in male rats. But only males. In female rats and mice of both sexes, no link was found.

What does this tell us? Mostly, that the jury's still out. On more than just tumors as well. Various hypotheses have postulated a link between EMFs and sleep disorders, autoimmune disease, certain cancers, anxiety and other mental health issues, general aches and pains, and headaches. As with allergies, some people are more sensitive than others. A small number of individuals appear to have hypersensitivity to EMFs.

So yes, we know without a doubt that there is potential for adverse health effects from EMFs. At extremely high levels, EMFs can even damage human cellular DNA. Luckily, the levels required for that damage are far higher than anything you'll encounter in *normal* society. But you should be aware that EMF levels are increasing with both the number of new IoT devices and increases in the deployment and strength of the cellular networks themselves.

5G: Typhoon or tempest in a teacup?

5G, if you're not aware, is the latest iteration of the network used by your cellphone, laptop, and other connected devices. Not surprisingly, 5G supersedes 4G which superseded 3G. Each 5G network is a cellular network where the service area is split up into small geographical areas called cells. Each cell contains a local antenna that 5G-compatible wireless devices use to connect to the Internet and telephone network by radio waves. How is this different from 4G? A few ways. First, 5G uses higher radio frequencies that allow it to carry more information at a faster rate. That increased bandwidth (i.e., the inclusion of higher frequencies) allows 5G networks to deliver faster download speeds—eventually up to 10 gigabits per second.

While the higher bands in 5G are faster and carry more information, they are also easily blocked by physical objects like buildings and trees. To get around that, 5G uses multiple input and output antennae to boost the signal across wireless networks. Those antennae include small transmitters placed on buildings and other objects plus a substantial number of satellites. Elon Musk's SpaceX alone will deploy 50,000 Starlink Satellites to create a global network (a Starlink "Constellation"). The FCC has already approved deployment of thousands of these satellites in very-low-earth orbit at altitudes just above 550 km (340 mi). Also approved are the installation of one million little-ground antennas. That's right, a *million* antennas just to get started. And SpaceX is just one company. What about Amazon, Google, Facebook? What happens when all the current Internet service and cell service providers invade that market? That Starlink Constellation could become just part of an awfully crowded Zodiac, all of which would be emitting high-frequency radio waves.

Also keep in mind that 5G superseded 4G, but it didn't replace it. All the 4G networks are still out there, still producing their own EMFs. In fact, some of the 3G technology is still out there, although beginning to lose functionality. That's really the issue with trying to assess the safety of 5G, or any of the EMF-producing technologies. They don't exist in isolation. No one knows what issues will result when you factor in existing EMF levels from prior technologies. Experts in many fields are questioning just that.

So what should you do? Just be aware of the potential issues. The Global System for Mobile Communications Association (GSMA) predicts 1.7 billion 5G network subscribers by 2025. 5G is clearly not going away. The rapid expansion of the Internet of Things requires faster, higher capacity networks. There are also other benefits. Starlink Constellations by their very nature are fast to deploy. They can be used to replace infrastructure destroyed by natural disasters. In February 2022, talks were underway to deploy a Starlink Constellation above Tonga after a volcanic eruption destroyed the undersea cables connecting that island nation to the Internet.

There are no clear-cut studies about the health risks of 5G but one thing is clear. There will be a LOT of satellites beaming down from space to antennas and transmitters to all kinds of devices in your home, outside, in your car, and well just about everywhere. By the end of 2021, there were already 31 billion IoT devices worldwide. By 2031, experts predict 125 billion devices will be in use. Cellphones, satellites, antennas, self-driving cars, billions of devices—can you say electromagnetic smog? I think you have the picture.

Until the scientific studies are in, it's unclear how 5G higher bands, antennas, and devices will affect us. It could take a decade or more before we know. In the meantime, we're seeing the interaction of billions of devices with human health. This certainly is the biggest technical human experiment in my lifetime. We have three choices:

1. Go live in cellphone-free zone.
2. Use the technology and don't worry about it.
3. Use the technology, but reduce your exposure to it.

Door number one is unlikely to open. While you might try to live in a cellphone-free zone, they're getting harder and harder to find on this planet. The goal of 5G is a global network relying on satellites to allow even the most remote and inhospitable areas to be networked.

Option number two has been our societal default. To a degree, even the scientific community has been rather mum about the potential hazards. That's not because individual scientists aren't concerned. It's more like they're concerned but don't have an answer yet so they're withholding comment while they research. The research is also a lot more complicated than I covered here. Even if there was more clear data on the effects of EMFs on human health, it would still be unclear to what extent EMFs alone are responsible. We live in a world that also contains pollution, pesticides, and other environmental toxins, not to mention the health risks in the processed foods many of us consume and the cumulative effects of poor lifestyle choices. If EMFs are one more risk factor in what have come to be seen as "diseases of progress," how can we possibly identify the extent of their responsibility? In the face of this complexity, a lot of people are choosing to simply ignore the potential hazards and just wait for the experts to decide. There are some benefits to this approach. It's easy. And it's possible that this really is a tempest in a teacup. On the other hand, it could be a cigarette. How long did it take the experts to acknowledge that smoking was potentially fatal? Sometimes waiting for the experts to decide is waiting too long.

Personally, I'm going for choice number three. I plan to use the technology for all the benefits it brings but take some simple precautionary measures just in case. For example, I'm on the waiting list for Elon Musk's Starlink service. I can't wait to get my hands on it! As I set it up though, I'll be looking at the best way to reduce my exposure inside my Airstream. For now? I mitigate my EMF exposure in several ways. I keep my router out of my sleeping area and turn it off while I'm asleep. When I'm not working, I often turn my cellphone off. At the very least, I keep it out of my back pocket. I also try to limit the electrosmog I personally generate with a regular digital detox. It's like a time out but applied to all technology. No phone, no laptop, no gaming systems, no TV… For even better results, I try to combine my digital detox with some forest bathing. Both disconnecting from technology and forest bathing have been found to have significant benefits to both physical and mental health. So even if it turns out that EMFs aren't a major health problem, I'm still ahead. It's a no-lose solution.

Where's 5G now?

Wondering if there are active 5G towers in your neighborhood? You can check the Ookla rollout map (*https://www.speedtest.net/ookla-5g-map*). This interactive map will tell you if there are 5G towers close by. (Be aware that you start at the global level. You may need to click a lot to zoom in on your town.) You can also check Verizon (https://www.verizon.com/coverage-map/) and T-Mobile 5G (*https://www.t-mobile.com/coverage/coverage-map*) coverage maps.

Can your Airstream protect you from EMFs?

Now that you know what EMFs are and why you might want to be concerned about them, you may be wondering why I brought that up in the first place. Good question! It turns out that if you're concerned about EMFs, the Airstream design and construction may help to allay some of those fears.

When Wally Byam designed that first silver bullet Airstream, no one was concerned about 5G or EMFs or health effects of technology. He chose aluminum alloy for the exterior because it was shiny, artistic, lightweight (so good for gas mileage), and possibly most of all, inexpensive. I bet Wally had no idea that his *artistic-retro* design would carry us into the future in a way he had not imagined. When I go into my Airstream and close the door, I'm in the world but at the same time I feel free from it. I know that sounds strange, but I have asked other streamers if they have a similar feeling and they said a big, yes. Some Airstreamers tell me it's like cocooning. So, what is it about this aluminum design that Wally didn't know back in the '30s? Well, something that most people know nothing about. That cocooning feeling of being protected may well have something to do with the aluminum that surrounds you.

Because of its aluminum construction, the Airstream exterior functions as a Faraday shield. A Faraday shield (sometimes called a Faraday cage) is an enclosure made of specific conductive materials that block electromagnetic fields. Faraday shields were invented by Michael Faraday in 1836. For almost 200 years now, they've been used to protect sensitive electronic equipment from excess EMFs.

How does it work? The shield's conducting material (the metal aluminum in the case of the Airstream) distributes the electrical charge across the shield surface, preventing the EMFs from reaching inside the shield. A good example of how this works is lightning striking a car. When lightning strikes, the metal car exterior functions as a Faraday shield, conducting the electrical charge along the outside of the car frame thus preventing the charge from entering the interior of the car.

Is it perfect? Definitely not. In our lightening example, simply sticking your arm out the window and touching the car frame brings the charge into the car interior. On the Airstream, there are gaps in the Faraday shield. Windows, windshields, etc. These gaps allow some EMFs to pass through. Which is good. If the Airstream blocked 100% of EMFs, your cellphone wouldn't work at all inside. Which would be bad. Still by functioning as a Faraday shield, your Airstream can limit your exposure. That is, as long as you don't overload your technology inside. You can also minimize your exposure further overnight by using EMF blocking curtains while you sleep to close the gaps.

Think of the Faraday shield function as an umbrella to keep out electrosmog. Using an umbrella doesn't guarantee that you'll never get wet—especially not in a serious downpour. But you will get a lot less wet than you would if you didn't have an umbrella. The Airstream Faraday effect won't prevent 100% of electrosmog from entering.

Measuring EMFs

One of the big issues with testing Radio Frequency (RF) levels (5G) is that the industry hasn't agreed on safe exposure limits. The IEEE has one measurement; the FCC has another. Russia and China have their own different safety limits.

While there is considerable debate over what level of EMFs are safe, the starting point is to measure the EMFs you're being exposed to in various locations and situations. This will give you a good feel for the RF signal strength so that you can determine your comfort level and look for ways to reduce your exposure if needed.

You can easily measure RF with an EMF testing device. RF is measured in milliwatts. A milliwatt is a measurement of power and is equal to one thousandth of a watt.

While professional in-home testing can run \$300–\$500, handheld testing devices can be purchased for well under \$200. Their efficacy varies considerably so be sure to read the reviews carefully.

I used Alpha Lab's Trifield (TF2) EMF reader to measure the RF signal strength inside and around my Airstream. The TF2 meter measures radio frequency bandwidths from 600 MHz to 6GHz. With my 5G hotspot turned on inside my Airstream, the RF measured highest at my desk (where I work). EMF levels at my desk measured at a peak of 12.3 milliwatts with all technology turned on to a low of .08 with all technology turned off. In my sleeping area, RF levels were significantly lower at 0.3 milliwatts regardless of whether my office area technology was turned on or off. Adding EMF protective shades, and with all technology off, RF levels dropped to .019.

I was very happy with the low levels in my sleeping area. That low level is important to me because I spend at least eight hours a night sleeping there. And this is important to people who are sensitive to EMFs. Think of it like an allergy. People who are sensitive to a specific substance—say pollen or gluten—can be miserable with levels that would have little to no effect on the rest of the population. The World Health Organization recognizes this issue as electromagnetic hypersensitivity (EHS).

My testing shows there are some simple things you can do to lower your EMF exposure:

- Turn off your technology when you're not using it.
- Know where your RF devices are and keep as many of them out of your Airstream as possible.
- If you need to keep your cellphone on, keep it out of your pocket and away from your sleeping area.

Despite being inconsistent between agencies, safety levels are based on what's safe for the general population—not individuals with EHS. You may need to determine your own safe level by measuring and reducing your exposure. If that sounds too complicated, you can always just turn off devices when they're not in use.

External factors

Like pollution smog, the level of electrosmog depends on where you are. In general, the more urban your environment, the higher exposure because you'll be surrounded by IoT devices as well as denser cell network coverage. But general isn't always true. You could be in the middle of nowhere and still have cell towers nearby, in which case your exposure could be higher. Like electrical transformers, cell towers have a bit of a bad boy reputation. City planners won't necessarily point them out in the tourist brochures. 5G is even more complicated since the smaller 5G antennas aren't always easy to spot. That's why it's important to test.

If you find the results higher than you're comfortable with, there are several steps you can take to lower exposure. First, check if there are cell towers or 5G antennas nearby. The further you get from the object sending out the EMFs, the more the intensity decreases.

On one trip, I ended up in Thermopolis, Wyoming. When I arrived, I had zero cell and zero Internet in that location. I had no idea that it was a dead zone until I arrived. I quickly decided that it was a perfect weekend for a digital detox. Even better—the Wyoming Whiskey distillery was right down the road. If I hadn't stumbled upon a low EMF area, I might have missed Wyoming Whiskey's

Distillers Select 112 proof, which is only available at the distillery gift shop!

If moving away from the EMF source isn't an option, you can also increase the effectiveness of the Faraday shield effect by installing EMF-blocking curtains. Tin foil hat jokes aside, you can actually use aluminum foil on the inside of windows as a low-budget EMF blocker.

Route Planning

I started camping with my grandparents when I was seven years old. Papa used to say, "Safety takes planning, and we plan to stay safe." He knew that thinking ahead was the key. We camped every single summer for the next decade together. Not once did we have a problem.

Safety planning for your Airstream adventure begins with selecting a safe route. Not surprising. After all, a safe road trip requires a safe road.

If you're like most people, you leave the route planning to your GPS. If I'm heading out in my Toyota Tundra, I do the same. And honestly, I used to do the same with my Airstream. I learned not to after an extremely long, stressful trip the first time I took my rig to Colorado. The highway I'd originally planned to take was closed for repair. The detour took me through the Colorado mountains along one windy mountain road after another—each filled with hairpin turns and not a guardrail in sight. It would have been a scary drive in just my truck. Towing a rig while maneuvering those endless hairpin turns was terrifying. I spent the entire three-hour drive praying instead of looking forward to getting there. When I finally arrived three hours later than expected, I was exhausted. It was one of the scariest drives I've ever taken. And needless. Had I checked the route online instead of relying on my GPS in real time, I would have discovered that the main road was closed, learned when it was scheduled to reopen, and had a chance to preview that detour. I would have easily seen the extreme curves and known upfront that it wasn't a safe drive towing an Airstream. Either discovery would have allowed me to plan a safer route.

While you're checking out routes online, don't forget to print a copy. Paper maps can be indispensable, especially in remote or mountainous areas where Internet connections can be spotty or non-existent. It's smart to have hardcopy maps for the places you're traveling to. At the very least, you'll want to keep a good atlas in your rig. Both National Geographic and Rand McNally make several detailed road atlases. My preference is the *Rand McNally Road Atlas and National Park Guide* because I especially enjoy touring national parks in my Airstream. There's also a trucker's guide with information on restricted roadways and laminated pages for durability. While that concept is nice, I prefer to just replace my Atlas annually. Even if I don't end up lost and cut off from the Internet, it's still well worth $25 a year just for day dreaming new trips in my rig. That's a great project to help kids (or some adults for that matter) learn how to read a map. You can start with the basics. It's surprising how few people realize that odd-numbered Interstate routes run north to south while even-numbered routes run east to west.

Weather Awareness

Twisty roads aren't the only things that can make your trip more hazardous. Sleet, ice, and pouring rain can impede visibility and make maneuvering more difficult—especially while towing a large Airstream. Imagine trying to corner on black ice with a 33-foot Airstream Classic. I get shivers up my spine at just the suggestion!

General weather patterns are mostly a given. Maine in February will nearly always be a frozen mess. Likewise, New Mexico in August is going to be stifling. Specific patterns though, especially in transitional months like March/April or November/December, can be harder to peg, especially with changing weather patterns. Some years, Thanksgiving is a winter paradise. Other years it's more a continuation of Indian summer. Because weather patterns change rapidly, you may not know whether you'll be planting flower bulbs or shoveling snow at week end.

When you're planning a trip around questionable weather, keep in mind that GPS arrival time is always wrong when you're towing a rig. Also factor in how many stops you plan and leave yourself sufficient time for decent meals. Since rigs are hardly drive-thru friendly, you're likely to spend a little longer on meal breaks. Knowing that, you may want to plan ahead to find interesting stops. I enjoy sampling local delicacies and regional dishes. If I'm in Maine, I want to sample the fiddleheads with my lobster. In Gettysburg, I want the Civil war inspired game pie, pickled watermelon rinds, and spoon bread. Beignets and cafe au lait in New Orleans. A habanero rattlesnake burger at Uncle Tony's Café in Spring, Texas. Okay, I'm not that adventurous. I'd take a pass on the rattlesnake. But I'd certainly enjoy sampling the wares of any chef bold enough to include that kind of burger on the menu. The Airstream lifestyle doesn't just let you get away from it all. Sometimes, it takes you places you wouldn't have thought to go. Be sure to keep an eye out for them.

Regardless of where you go or how often you stop, you always need to account for weather conditions. I love getting on the road knowing that not one drop of rain is forecasted. However, I know people who get on the road without ever looking at the weather. That's crazy, especially given our unpredictable weather patterns lately. For example, it was below freezing here in Austin for the last two days. It was clear that a storm was moving across the country and that Austin would be below freezing for two days. A serious ice storm was expected. No surprise. Yet, I spoke to a guy who just pulled in the night after this ice storm began. He told me it took him two days to drive six hours. I think he must have forgot to look at the weather. That's dangerous not just for you but for other people on the road too.

If it looks like a big storm is on my route, I may delay my trip for a day, if necessary. If you are driving through areas that are known for tornados, it's important to understand what time of the year these storms come through and if they might threaten your trip. The last time I drove though Kentucky, a big storm was heading that way. Because I checked the weather in advance, I left a few days early to stay ahead of the storm. That was good planning because two days after I drove through, five tornados touched down and devastated Kentucky and Tennessee. Tornados can touch down at unpredictable times of the year too, so it's a good idea to have your cellphone turned on for tornado warnings and weather alerts.

When I'm staying in my rig overnight or long-term, I have an Eaton weather radio that has a weather-alert feature that will turn the radio on and warn me of dangerous unplanned weather. If the weather changes while I'm sleeping, the radio will turn on and alert me of the potential danger. It has rechargeable batteries that last a long time before I need to recharge them, as well as a solar panel. I like having the weather radio because anything can change overnight, and this warning feature could save lives.

Site Selection

Selecting your campsite is a critical part of trip planning, especially if you're planning an extended stay. Because RV travel has increased significantly in the past few years, it's also important to do this well in advance of your trip. To find a site in some areas, such as in or around popular national parks, you may need to reserve as much as six months in advance.

Deciding where to stay is as important to your safety as selecting a safe route to get there. Because I often travel alone, I like to stay at well-established campsites with plenty of other campers around. Not surprising, I often end up at well-known chains like KOA, Sun Outdoors, Good Sam, etc. They usually have good lighting, solid security, and clean bathrooms. Some campgrounds have security gates and security codes required for entry.

I also check to see what amenities they have. While safety comes first, it's nice when it comes along with a pool, gym, and dog park. Other amenities are less obvious. When I stayed in Austin Texas during the winter, I loved staying at the Lone Star because they delivered propane right to my site. When the winter chill set in and I needed to change the propane, that door-to-door propane service became priceless.

> **NOTE:** RV theft is much more likely in some places than others. The states with the highest levels of RV theft are Kentucky, West Virginia, and Florida.

While a well-known name can be a good indicator of quality, it isn't always a guarantee. To be safe, do your homework. Before booking your reservation, check the reviews on a site like *camgroundreviews.com*.

Also don't be afraid to admit your mistakes. Even after doing my research, I've occasionally shown up to find the place that I'd reserved just didn't make me comfortable. If a campsite makes you uncomfortable, you should move on. It doesn't matter exactly why. The site may come across as too secluded, the bath house too grimy, or maybe you're just not comfortable with the behavior of other campers. Whatever the reason, trust your Spidey-sense. And remember that you don't have to stay.

One time this happened to me, I walked into the campground office and said, "This isn't what I was expecting from your website. I'm just not comfortable staying here." They cancelled my reservation, refunded my credit card, and I found a great place where I *was* comfortable just 30 minutes down the road.

Ask Yourself

- Does my cellphone really need to be in my back pocket ALL the time?
- Am I tied to my cellphone when I really don't need to be? Can I set it down and turn it off when I'm focused on other tasks?
- Does my router need to be on at night? Can I limit my EMF exposure by turning it off overnight?
- Should I schedule a weekly digital detox? Can I limit my EMF exposure and promote mental well-being by choosing a regular day and time to go "technology free"?
- When did I last test the amount of radiation in my Airstream? Was I happy with the level?
- Is there a current road atlas in my rig?
- Do I have a weather radio? Should I get one?

- Have I planned a safe route to my destination?
- Have I checked the weather reports along that route?
- Is a safe, comfortable camping site reserved at my destination?
- Is the site I chose near known cell towers or 5G antennas?

4

PREPPING FOR JOY

Over the past few decades, the concept of prepping has gotten quite a negative connotation. Be honest. Doesn't just the term "prepper" invoke images of tin-foil hats and bearded harbingers of doom—stockpiling canned meats and military rations, awaiting the end of the world?

It's ironic, isn't it? In reality, "prepping" as in "preparing for possible problems" is the best way to ensure that bad doesn't happen. Being properly prepared will help you to avoid many major problems and comfortably ride out the rest. It also puts you in a place mentally where you're able to choose a destination of joy, regardless of any bumps on the way to get there.

Airstream founder Wally Byam understood the concept of being prepared to choose joy when he set out to make the best possible travel trailer on the market. For close to a century, his team delivered a fully contained rig with everything you need to live. I don't think Wally believed that being miserable in an Airstream should be part of the package. As one of my fellow Streamers pointed out to me, "I chose to live in my Airstream as a lifestyle choice. Being miserable at any point just isn't a part of that choice."

Being miserable isn't an option for me either. I choose joy. You're probably thinking, "But stuff happens." And it does. When you're unprepared, the little curve balls that life throws you can very quickly spiral out of control. But when you're prepared for the possibilities, choosing joy can become your default move.

Consider a power outage. Sitting in the dark in cold weather with nothing to do is just asking to be miserable. Cuddling in comfy jammies with extra blankets for an evening of family games by candlelight to the sounds of the forest or ocean, on the other hand, is the definition of relaxing. The difference? Planning ahead.

When I got my Airstream, my goal was to be as self-sufficient as possible. Wally made the Airstream fully contained but being self-sufficient is totally different. Jokingly, I used to call preppers, the tin-foil hat crowd. This is what I say now, "The preppers were right."

I lived in California most of my life and I never prepared for an earthquake or any kind of disaster. I had enough food and water in the house for a few days at most. Thank goodness a real disaster didn't strike.

Since I adopted the Airstream lifestyle, I've changed my thinking. There's something about carrying my home with me from place to place that seems to make me more conscious of my responsibility for protecting it. These days, I plan ahead. I never let my gas task get under half full. As soon as I got my new Airstream, I made a list of all the things that I might need in an emergency. I prioritized my list and started chipping away, one item at a time. In short, I began planning for the unknown.

Clearly, you can't plan for every unknown problem. But you can have all the likely ones covered well in advance. The most expected of the unexpected? That would be vehicle breakdowns, power outages, extreme weather, and being off-grid without access to amenities (including groceries and restaurants).

Vehicle Breakdowns

The more time you spend on the road, the more likely it is that sooner or later you're going to have to deal with some type of vehicle breakdown. Add in the fact that you're towing the equivalent of a small house behind you and those odds jump to nearly 100%. You can smooth over this problem well in advance:

- Choose an adequate towing vehicle.
- Invest in comprehensive roadside assistance.
- Pack the necessary emergency supplies.
- Be cautious of creepy people and check your lug nuts.

Towing vehicles

If you're buying new, your Airstream dealer can provide detailed information on the type and capacity of vehicle you'll need to tow the Airstream you've selected. Obviously, that will depend on the rig you select. If you're heading high-end for a 33-foot Airstream Classic, you'll need more power in your towing vehicle than you would for an entry-level 16-foot Basecamp.

Of course, you may not be buying new. Remember, 70% of all Airstreams ever produced are still in use. That means a large and growing secondhand market. If you're buying used, you'll need to do your homework. Whatever you do, don't just assume that because your tow vehicle is somewhat largish that it's also powerful enough to haul a travel trailer. Just ask my editor about that. A few years back, her oldest daughter spent a gap year traveling with a Renaissance Faire—a perfect occupation for RV living. When she purchased a used RV, she just assumed that her parents' old minivan would have no difficulties towing it. Not a great assumption! Three states later she still had the used RV, but it was parked beside an undrivable minivan with a totally trashed transmission.

Roadside assistance

So far (knock on wood), I've only broken down once. But it was certainly a learning experience. I was towing my Airstream with my Toyota Tundra to Florida when I had a problem with a tire. Not a flat tire. That would have been annoying, but within my ability to fix on my own. And honestly, the annoyance really would have been balanced by my pride in having learned enough basic maintenance to take care of a flat tire on my own. This was more complicated. One of the tires on my trailer came off altogether while I was driving. That pushed the problem from annoyance to major safety concern.

It was pretty clear I needed a tow to a reputable garage for a repair. No big deal, I thought. After all, I'd been paying for roadside service for 30 years just in case of a problem like this. I made the call, thoroughly confident that my years of roadside service payments would be rewarded with fast, friendly service.

Yeah. I don't know what I was thinking either. The time estimates I was given (each of the many times I called) were in hours, not minutes. Overall, it took 23 hours from my call for help until the tow truck actually arrived.

During those 23 hours, I was at risk on several levels. First, the location was awful. I was just north of Tampa, Florida on Hwy 75. While I was able to pull to the side of the freeway, I wasn't able to continue to a rest stop or even a wider berm area. Instead, I was barely out of the slow lane on the road for almost a full day as trucks zoomed by me at 75 to 85 miles an hour. Because the location was so awful, I wasn't able to safely get into my travel trailer for food, water, or supplies. Just to add a layer of scary to the baseline annoyance and frustration, I was far enough between exits, and away from good lighting, that I was genuinely afraid that some stranger would stop by after seeing a lone woman broken down. Those cheerful helpers might seem a godsend in broad daylight in the public area of a populated rest stop. On the side of a four-lane highway in the dark, I'd be less excited for the potential help and more likely to debate, "Good samaritan? Or serial killer?" After midnight, they all kind of look the same.

So how can you prepare for an unexpected breakdown? First, make sure your towing vehicle has regular detailed maintenance. In particular, don't write off the minor movements that don't feel right—an ever so slight wobble at high speed, or even a light pull on the steering at sharp turns. Pulling an RV is a heavy-duty task for your vehicle. Treat it like a star athlete in routine care and preventative maintenance.

Top-notch care will prevent some, but not all, possible breakdowns. After all, I did give my Tundra top-notch care and was still sidelined. Well cared for cars may last longer, but well-used vehicles still wear out faster. And sometime accidents just aren't preventable.

Adding more maintenance wouldn't have prevented my breakdown. But having better roadside service would have made the breakdown easier to deal with. When I began streaming around the country, I kept the same roadside assistance program that I'd relied on while driving a few miles each week between home, work, and shopping. That wasn't the best idea.

Roadside service isn't a one-company deal anymore. There are several high-profile independent roadside service providers. AAA and On-Star immediately come to mind, but they're only two of a dozen or so nationwide including Good Sam, Better World Club, and

CARCHEX. In addition to the independent providers, there are roadside add-ons—sometimes for as little as the cost of fast food—from many of the major insurance companies. A few are limited to specific groups. AARP offers roadside assistance for seniors. USAA for current and former military. And there are free options still included with some credit card offers. If I'd thought of it at the time, I could have used the free towing provided with my American Express card. They may or may not have provided faster service.

Another factor to consider is that some roadside assistance services have better coverage in some areas than others. Look for regional coverage that matches where you're most likely to travel. If you're planning to explore the Pacific Northwest, you probably don't want a service that focuses on the Midwest or New England. If you're travelling cross country, or just often, consider doubling up on services. Most cost under $100 a year. A second service could double your chances of getting prompt service when you need it. Even with two services, you'd still be paying less than the cost of a single average tow.

Emergency supplies

Remember the example I gave earlier about the power outage? And how having the right supplies turns a miserable evening into an opportunity to relax? The same concept applies to breakdowns.

Preparing for possible vehicle breakdowns requires pretty much the same preparations you'd make for any trip with your trailer. The important part is that these items need to be in your towing vehicle—not in your rig. Don't assume you'll be able to safely get into your rig to retrieve any items that you'd need. When I broke down, I was far too close to a highway of speeding trucks to safely access anything in my Airstream. Luckily, I'd thought ahead and placed the following items in my towing vehicle:

- Water
- Food
- Warm clothes
- Blanket and sleeping bag

- Flares
- Reflecting cones or triangles
- Flashlights
- Battery phone charger and power bank
- First-aid kit
- Spare keys
- Portable tire inflator
- Battery jump starter

Food and water

Because I'm not a fan of fast food, I always keep a Yeti cooler with some healthy food options in the cab of my truck, just in case. I like the Yeti because it'll keep even ice frozen for three to five days in normal use. Because of that, when I broke down off of Hwy 75, I was able to have a fresh salad, mixed nuts, and some berries along with ice cold water. It wasn't a gourmet meal, but it held me over while I waited for roadside assistance. If you prefer comfort eating in the cab, you can replace that fresh salad and berries with chips and salsa, Doritos, pretzels, or cookies. (If you go for salty though, you'll want to up that 6-pack of water to a 12-pack.)

The amount of emergency food and the level of cooler you opt for will also depend on how you travel. Friends of mine once did a cross-country drive over the course of an entire summer. Each day, they broke camp at 6 a.m. and drove no more than three to four hours. Even with breakfast, they got where they were going just in time for museums, galleries, and shops to open. Given their streaming style, they're fine with an extra meal or two of snacks and a $20 cooler. I travel a lot differently! I prefer to get my driving over within one or two days then stay put for a week or so, really delving deep to explore each destination. I'm also very much enjoying going off-grid, selecting fairly remote destinations where I'm likely to find fewer amenities a lot further apart. For my streaming style, I like at least five days of emergency rations (plus a healthy fresh option or two). Because I'm seriously attached to my salads, I'm okay with paying for a premium cooler. And I keep a full case of bottled water in the cab, just in case.

Warm clothes and blankets

Warm clothes and blankets are often a must even if you aren't broken down. Remote mountainous areas are surprisingly cold at night, even in the height of summer. Consider Arizona. The pine forested central areas that sport normal temps in the 80s on summer days still drop as low as 35 degrees overnight. California can also be surprisingly cold, especially to East Coast Streamers expecting warm balmy weather. Sequoia National Park posts temps below zero overnight in winter. I honestly did not realize that on my first camping trip there. I ended up wearing every piece of clothing I'd brought simultaneously every time I was outside. Proper cold weather clothes would have been much more practical, and comfortable. Today, that wouldn't happen. I always leave a spare hoodie and an extra pair of sweat pants in the truck—just in case!

If your towing vehicle does break down, you may or may not still have heat in the cab. For basic comfort, be sure to bring along a blanket or sleeping bag for each person. A lot of people go for the Mylar thermal blankets, sometimes called space blankets. The plastic composite sheets are inexpensive, lightweight, and effective. They'll also retain up to 90% of the user's body heat, so they're pretty effective in an actual emergency. And they're available by the 6-pack for as little as $20.

While I love the aesthetic of the space blankets—very *Star Trek*—I much prefer the feel of my mummy wrap sleeping bag. If you prefer a sleeping bag as well, be sure to check the temperature rating and compare it to the overnight temps where you plan to travel. Most commercial sleeping bags are rated for 20 to 40 degrees. That's fine for summer overnights in most areas. However, if you're planning a lot of high-altitude camping, or winter streaming in colder areas like New England or parts of Canada, you'll want a warmer sleeping bag.

In general, there are three levels of sleeping bags. Summer (or low-altitude) sleeping bags are perfect when the weather is above freezing (32 degrees). Three-season bags, intended for spring, summer, and fall, are rated for 20 to 32 degrees. Below that are the winter/high-altitude bags. For guaranteed comfort, you'll want to check the ratings. While ISO (the International Standard) is the rating most commonly given, you may also see EN ratings (European Norm). EN ratings usually provide two numbers. EN Comfort is the comfortable sleeping temperature while EN Lower Limit is the lowest temperature at which most people would be able to sleep without waking up because they're too cold. For the truly adventurous, there is a third number—the EN Extreme. That Extreme number is the lowest temperature at which the average woman could spend six hours in the bag without dying of hypothermia. That's a little too extreme for me. I selected a sleeping bag that provides a decent EN Comfort rating for three-season camping. Truthfully, if there'd be ice forming on the windshield, I'm more likely to reschedule my trip for better weather.

Whichever type of sleeping bag you select, remember that this needs to fit in the cab of your rig—not packed away in the trailer, which you might not be able to get into in an emergency. Also keep in mind that a sleeping bag will take up a lot more space than a Mylar space blanket. For me, that works just fine. If you're bringing the entire family, you'll probably want to just pick up a 6-pack of space blankets. Properly packaged, most Mylar blankets take up little more space than a package of microwave popcorn, making them very easy to just tuck under the seat or in the glove box.

Flares and reflectors

My first move when I broke down was putting on my truck's emergency flashers and setting my flashlight to blinking mode. The combination of flashers and that blinking flashlight angled out the window helped speeding truckers to notice me on the side of the road.

Thankfully, I was able to pull onto the side of the road. Had I broken down in the center lane and been unable to move, I would also have set off a road flare to warn motorists behind me. Since road flares burn relatively quickly, it's good practice to carry several.

Both flashers and flares are short-term solutions. Road flares last on average 10 to 30 minutes each. Flashers last for the duration of your battery. This is good news if you run out of fuel. Bad news if your problem is a dead battery.

For a power-independent solution, you should also carry reflectors. I carry a plastic triangle reflector in my truck, stored next to the flares. It's very similar to the triangle reflectors posted on the back of Amish buggies so the horse-drawn carriages aren't run off the road by unobservant motorists. Ideally, you should carry three reflective triangles. Place one directly behind your vehicle, the second 10 feet behind, and the third approximately 100 feet behind. Most reflective triangles simply reflect but some are equipped with battery powered LED lights as well.

Flashlights

Your flashlight is your most basic piece of emergency equipment. After all, nothing makes a scary situation worse than sitting in the dark. I really genuinely hate sitting in the dark, so I make sure to have three flashlights available at all times. A battery flashlight (with spare batteries), a solar flashlight (that's easy to charge), and a keychain flashlight so I've got an easy light source I won't lose and can use to find the other flashlights in the dark. Just in case I can't find them right away, my keychain flashlight is rated at 1000 lumens and does a great job of illuminating my immediate area.

If you're wondering what a lumen is, it's a measure of how bright a light is. In the past, we bought light bulbs based on how much electricity they used. If you're over 30, you no doubt have a good grasp of the difference in light produced between a 40-watt bulb and a 100-watt bulb. Those were great measurements for incandescent light bulbs, but they really don't translate to LEDs. LED lights use around 75% less electricity. An LED light drawing 100 watts would light up the block, not just the inside of your rig. For this reason, light bulbs and flashlights are now described in terms of how much light they produce instead of how much energy they use. A traditional 40W bulb produces 450 lumens, a 60W bulb produces 800 lumens, a 75W bulb 1100 lumens, and a 100W bulb 1600 lumens.

Battery phone charger

The only thing worse than sitting in the dark is feeling cut off from the rest of the world. Your cellphone is a critical part of your safety kit.

When I broke down, I depended on my cellphone. Obviously, it put me in touch with the towing company that (eventually) towed my rig to safety. Less obviously, my phone kept me safely in contact with others.

I might have been sitting alone with my broken rig on the roadside, but I wasn't out of touch. The first thing I did when I broke down was to call my brother. He helped me calm down and focus on getting a tow. I also called a friend who I had been talking with earlier that day. He also helped me stay calm. He called me back later that night and stayed on the phone with me for a few hours. I made sure he knew exactly where I was. I've always had this deep-rooted fear that I would get stuck on the side of the road and a stranger would stop. That's exactly what happened. I didn't say a word to the guy and he eventually went back to his car. Just being on the phone with my friend helped keep me stay calm and it may have frightened off the guy simply by making it clear that I wasn't "really" alone. And yes, I realized that the odds of a guy stopping by while I was broken down were incredibly low. At the same time, that's exactly what happened. I can't convey how much safer I *felt* being on the phone with my friend. I'm sure my friend had a million other things to do that night, but he helped me make it through a few hours of what turned out to be a very scary, exhausting night.

Your peace of mind, like mine, is absolutely worth a decent phone charger. Because you don't know if the car battery will be working, I strongly suggest getting a power bank in addition to leaving a phone charging cable in the cab.

I know that my ordeal sounds like a miserable situation. But in many ways, I was lucky. I was able to pull to the side of the road. I had plenty of water and some food. I had a blanket, a working cellphone, and reflective triangles. I even had a friend on speed dial to help me get through the night. In the morning, I found a restaurant a half mile down the road and got a cup of tea. That cup of tea was a true gift after a long night. It could have been much so much worse than it was. I learned from this experience, and it helped me become better prepared.

First-aid kit and spare keys

Every rig should be equipped with a first-aid kit. You can assemble your own, but it's easier to simply buy one prepackaged with basic first-aid supplies. Ideally, you should have one in the cab for any injury that occurs while you're driving or at rest stops. Before each trip, check your first-aid kit for any supplies that need refilled. Also make sure that you've included supplies of any prescription medications you'll need to take on your trip.

The first-aid kit in your towing vehicle is also a good place to stash a set of spare keys to your RV. If you keep a second first-aid kit in the trailer (a good practice!) you can be consistent and keep the spare keys to your towing vehicle inside that first-aid kit.

Why the spares? Few things are more frustrating than being locked out of your own home. In a traditional neighborhood, you'd park another set of keys with a neighbor. Or two. In the Airstream culture, your "neighbors" are in motion just as you are. While it's helpful to leave spare keys with a stationary friend, you really don't want to be stuck waiting for that friend to FedEx you those keys if you end up locked out.

Creepy people and lug nuts

Airstreaming full-time is a new adventure for me, but camping isn't. When I was a kid, I camped often with my grandpa. Part of my "job" was to help Papa get the camper ready for the road. When we packed the camper, Papa made sure that the weight was distributed evenly. We checked the lights to make sure the blinkers and brake lights worked. Last, we checked the tires on both the tow vehicle and the trailer. Papa showed me how to check the air in the tires and we always walked around the vehicles right before we left to make sure the tires still looked good. When we stopped for gas, he'd have another look at the hitch. Then he'd say, "Linda, walk around both sides and make sure the tires look good." I'd say, "We just did that before we left." And he'd remind me, "Every time we stop, we should check the tires." And we did.

I've carried that need to check and double-check the tires into adulthood. Every stop. Every time. So imagine my shock when one of the tires on my old Airstream came off while I was driving down the highway. The lug nuts came off and the tire just flew off. Luckily, I was in the slow lane on the freeway and was able to get to the side of the road. The speed limit on that freeway was 75 and I was in the slow lane going a lot slower than that. Another thing Papa taught me was that slow and steady gets you there alive. I can't recall him ever speeding, although I see other RVs zooming by me at 75 or 80 all the time. I know that going slow and being in the slow lane saved my life that day.

For months after this experience, I wracked my brain trying to figure out how those lug nuts could have come off. In online searches, Airstream forums, and many conversations with fellow campers, it became clear that this wasn't a common issue. Lug nuts normally didn't loosen themselves.

Then I remembered… The day before my accident, I'd had a run in with a very aggressive fellow camper. Because most fellow campers are genuinely nice people, the exchange really stood out. I was in Louisiana and had gotten tied up in unexpected traffic. It had taken an extra three hours to get to my campsite that day and I was exhausted. When I arrived, the office was closed so I checked the directions to my site in the confirmation email. Clear as mud. I had to drive around the campground several times to find my site. En route, I heard someone shouting as I drove by, but I assumed they were calling to someone else.

While I was unhooking my rig and setting up, a man came up to me complaining about my driving. His tone was somewhere between Rodney Dangerfield and *Mean Girls*. "Excuuuuse me, *Ma'am*. The speed limit here is 5 miles an hour. That's a rule and it's for everyone. You can't go 10 miles an hour how you were plowing through here. You need to follow the same rules as everyone else..." I'm sure you know the type. If we hadn't all been parked, I would have called him Mr. Road Rage. I did my best to pacify him. "Sorry. Thanks for letting me know. It won't happen again." He really didn't want to hear it. He seemed pretty determined to just stay angry. When I left him with his scowls to go inside (away from him), his last question was "How long are you staying? Are you leaving tomorrow?"

The next day, five hours after I left the campground with Mr. Parked Road Rage, my tire came off. Did I hear him skulking about, loosening my lug nuts in the middle of the night? No. Do I think it was a little bit too coincidental for comfort? Absolutely. Could I have prevented his (possible) vandalism? Probably not. But I could have checked my lug nuts in the morning before I drove off.

For details on the maximum mileage points at which you should check your lug nuts, you'll want to refer to the Airstream maintenance manual for your specific trailer. Personally though, I'll be checking my lug nuts every time I leave a campground with or without creepy fellow campers. It's a fast, easy check that I now know can save lives. It's also a good idea to carry a portable tire inflator.

Utilities and Power Outages

When you own a house, you are responsible for paying the taxes and local services. Your electric, water, sewer, and garbage are all services you pay for and expect to have. They just mysteriously work. You flip the switch, and the light turns on. You turn on the stove and the gas burners ignite. You turn on the faucet and you have water. These things simply work—until they don't.

Before I left California, I learned firsthand how easily these utilities can be shut off in a heartbeat. In mid-summer 2020, we hit a heat spell with high winds. In its midst, my electric company decided to just "take a little break" until the heat and wind subsided. With almost no warning, I lost electricity for a week. Crazy, huh? And no, I was absolutely not prepared.

With the exception of a few stalwarts and their backup generators, most homeowners are woefully unprepared for a power outage. Or water outage. Or really *any* short-term disruption in their utilities.

By their very nature, Streamers are much more prepared. For one thing, while most of us do connect to the power grid and utilities at campsites, we don't actually have to. RVs work equally well off-grid, so long as they periodically connect to the grid for refills.

Water and sewage

Water and its companion waste water (i.e., sewage) are great utilities for demonstrating the need for preparedness. I have to be honest, the one thing I was most freaked out about when I first started RV camping was the gray and black water hookup. It turns out not to be as big a deal as I thought. Now I compare the minor annoyance of flushing the black tank with the major annoyance of paying a $2,600 annual sewer tax in California just to flush my toilet. Overall, I prefer taking care of it myself now and keeping the $2,600.

Skipping the tax isn't the only major difference though. Let's consider two siblings: George and Gina. At George's condo, he turns on the bathroom tap to brush his teeth and clean water just appears. When he spits out the final rinse, the dirty water just disappears. Outside of paying his water/sewage bills and taxes, that's it.

George can be completely oblivious to both the source and destination of his water. Gina can't. She lives in her Airstream year-round, travelling the country. She loves the built-in showering capacity. The 45-gallon water tank goes a long ways—but not from California to Maine. If she wants to take showers once she hits the Midwest, she'll need to fill up her fresh water tank. To keep the water safe and sanitary, she'll also need to clean that tank at least once a year. Likewise, she'll need to empty the black tank (sewage) and grey water (used) tanks and flush them.

The big difference here is that George can be completely and totally clueless and the water will still turn on. Gina needs to pay attention.

She needs to know where her water comes from and where it goes and understand the system enough for routine maintenance and cleaning. She will need to pick the right water filter and hook it up at each location. She's a conscious consumer. George doesn't even need to be conscious—he can just consume. I much prefer being an informed and conscious consumer.

Electric

I always connect a surge protector to my outside power source. This helps to protect my electronics from voltage surges and spikes. For further protection, I installed an EMP shield to protect my truck and Airstream in case of solar flare, power surge, or electromagnetic pulse (*empshield.com*). Do we need to worry about solar flares and an electromagnetic pulse? I hope not, but I've taken precautions to protect my equipment just in case.

My Airstream has a desk and chair so that I can work from anywhere. The desk raises up so that I can stand and work. The Airstream has plenty of 110V outlets and USB ports at the desk area and on the bedroom side table. Of course, those outlets only work when I have power. Because I depend on having electricity to complete my work, I planned ahead to make sure I'd have power, even in the event of a power outage, or if I was in a remote location without access to the power grid.

My first line of defense for off-grid power is hardwired into my Airstream. When I bought my Airstream, I configured it for an off-grid emergency. I upgraded the solar to four panels and added three lithium batteries. My Victron phone app lets me turn on the lithium batteries and switch to solar for recharging. My app lets me see how many watts I'm consuming and how much power I have left. The batteries are bolted down inside the Airstream and the solar panels are mounted on top. I had an approved Airstream service center do the installation so that I didn't void my warranty.

I also have off-grid power that isn't connected to my Airstream. I purchased a lithium battery (Goal Zero's 1500x) and an additional 400 watts of solar panels. The lithium battery isn't sufficient to power my rig, but it can power lights, induction plates, space heaters, cellphone chargers, and computers. Since none of these additional power sources are hardwired, I can move the solar panels and the lithium battery anywhere.

I also purchased a backup generator. My friend did great research on generators. After I reviewed his top choice, I decided to buy the one he recommended (Honda EU2200i). I converted it to propane because I think it's easier to deal with than gasoline. I also have extra full propane cans set aside in case I need the generator.

Why do I have redundant off-grid solutions? If something happens to my internal off-grid power, I can still get power from my mobile off-grid set up. Also, when I prep, I think about helping other people. Just like the song lyrics go, *Think of your fellow man; Lend him a helping hand; Put a little love in your heart…* If I don't need the generator, and my buddy in Texas does, I can give it to him. If I don't need my mobile solar off-grid set up, I can lend it to someone in trouble. Helping other people in the time of disaster is where I want to be rather than wondering how I'm going to survive.

Extreme Weather

Over the past few years, I've learned a lot about extreme weather. Atmospheric rivers. Snow bombs. Fallstreak holes. Ball lightning. Thundersnow. Microburst storms. Extreme highs. Extreme lows. It seems like each new season brings some new and bizarre form of extreme weather I'd never heard of. In an era of growing weather instability, preparing for extreme weather takes on new importance.

We've already talked about some of the steps you'll need to take to prepare for the cold. Winter sleeping bags. Backup power supplies. Extra food. I'm super spoiled from living in California most of my life. I never thought I would stay in my Airstream when it dropped below freezing. I gave myself permission to go a hotel. I had the hotel picked out and my plan was in place. When the freezing rainstorm hit Austin in February 2022 and it got below 20 degrees, I figured hoteling was on the horizon. As the chill factor dropped below 20, the onboard furnace worked so well that this California girl did not need to go hoteling. Even in those cold temperatures, my Airstream kept me warm and cozy. My rig has two 30-gallon propane tanks and I found it easy to swap one out and keep the other one running. I spent $120 for propane from September to February. That's $120 for six months of heating. I used to pay $100 a month in California just to heat my hot tub in the winter.

You'll also need to prepare for the heat. Thankfully while the Airstream might look like a tin-can on wheels, bringing on visions of sweltering inside, the reality is much cooler. Cooler visually, of course, since no other RV manufacturer even approaches the level of style and quality in an Airstream. But also cooler literally, thanks to excellent air conditioning.

Even the older Airstreams provide excellent cooling, although not as inconspicuously as the latest models. I spent my first Texas summer in an older Airstream. The air conditioner kept me cool—even for a Texas summer—but it also made a lot of noise. I needed to replace the AC but my new rig was on order and I had to wait for it to be built. I was very glad to kiss that old AC unit goodbye for the new improved model.

My new Airstream has a climate control system that cools the cabin without being noisy. I actually have two air conditioners. My rig came standard with one, but I added the second one just in case one stops working. So far, I find that one AC unit keeps the entire rig cool, even with me setting that wall thermostat to fairly frigid. Still, I appreciate having that second unit for back up. That way, in the midst of a summer heat wave, I wouldn't find myself even temporarily without AC.

Surviving Off-Grid

While I'm traveling, I like spending time thinking about what's over the next hill rather than worrying about what I might need for my rig once I arrive. The Airstream's self-contained style allows me to focus on my journey and adventure. Since Airstream has been building since the '30s, they had the first hot water system, the first flushable toilet with a waste holding tank, and the first pressurized water system. Since then, they've had decades to perfect the internal systems. They really try to think of everything.

I built a house from the ground up, so I know what it takes to do things right to get superior results. You can build a solid house with the best materials, or you can make it look good and use cheap materials hidden behind the walls and floors. The same goes for travel trailers. Airstream not only focused on excellence on the exterior, but they carried that engineering focus to the inside too. Just like a well-built home, airplane, or private yacht, the Airstream is designed with engineering excellence. The moment I walked into my new Airstream I could see that engineering excellence carried throughout. The flooring, cabinetry, countertops, and appliances were all top of the line.

The reason you want a quality build is because you don't want internal systems breaking down while you're on the road—and especially not if you're traveling remote and off the grid. That means having all the equipment you need to survive and having all the supplies you need to thrive. In addition to the power supplies, water and sewage tanks, and heating and cooling systems, most Airstreams provide excellent kitchen and bathroom facilities.

Kitchen

While the kitchen is small (think efficient), it's also incredibly well equipped. Cooking is easy inside because of the convection oven which itself comes equipped with both broiler and microwave functions. The gas stove is great for frying and sautéing. Outside of a personal chef, which would be great, there's little you'd need to add. For me, I've added only one extra device to my kitchen, a portable electric induction burner that's great for stir frying veggies. Like any functional kitchen, my Airstream kitchen includes a refrigerator (with plenty of room for fresh and frozen food), a sink large enough for washing dishes, and locking cabinets (with hidden magnets) to help keep things secured while traveling. There are plenty of cabinets built in for pots, dishes, and food. I have enough food in the cabinets for at least two months. And I have extra storage space under the bed where I store extra food in case of an emergency.

While the older Airstream refrigerators used propane, newer models are electric. If you purchased an older Airstream, be sure to check your propane supply before trips and on a regular schedule while you're on the road.

In and out

Inside my Airstream, my kitchen includes leather benches and a big kitchen table allowing me to dine inside in luxury. It's a new guilty pleasure. At my brick-and-mortar home in California, I almost never ate at my kitchen table. The dry, balmy California weather was perfect for eating outside year-round.

I can't really do that at the height of summer in my favorite Texas RV park. Or during the beautiful (but wet) morning haze of a fall Seattle morning. Or a brisk and windy evening outside Chicago. On hot, cold, windy, or rainy days I can dine inside on my leather benches while appreciating the spacious table in my home on wheels. And if I have guests, that same table and luxury benches drop down into another sleeping area just like in the older Airstream models.

While the weather isn't always comfortable for dining outside, I do take the opportunity to dine al fresco whenever the weather does accommodate. Because the Airstream kitchen is designed to be efficient but small, eating outdoors is always the best option when entertaining guests.

Bathroom

The toilet and shower are in two separate rooms. I love this feature. I can't imagine families or even couples traveling without this great layout. The bathroom area also provides plenty of space for everything I need. A built-in mirror is mounted above bathroom cabinets that provide enough space for my hairdryer, hair tools, and makeup. Everything I need to access daily fits easily.

Like Wally Byam used to say, "We should not make changes—only improvements." Obviously, the Airstream team continues today to focus on improvements that make this fully self-contained rig just what you need when you need it.

Backup food

During an emergency, redundant cooking is just as important to have as redundant power. Even during a power outage, I have several cooking methods. I can use my electric induction burner connected to my portable lithium battery. I could also use my propane camp stove, or BBQ. If you're prepared with backup power, and made sure to stock propane cannisters, cooking is a breeze. Assuming that you have food to cook with!

How much backup food you'll need really depends on how and where you travel. If you're sticking to tourist areas and enjoying the commercial attractions, you can really get by with just a few days supplies. The odds that every store and restaurant in Disney World or Niagara Falls will close is pretty slim. If your goal is exploring the wilderness, that's a much different story. During their pandemic shutdown, some of my Australian friends took advantage of their time off to explore the Outback. Driving an entire family from Sydney to Darwin across the length of the Northern Territory required a great deal of backup food and water, as well as emergency petrol (gasoline). Chances are that your adventure is somewhere in between.

Cash

Most city dwellers have adjusted to a more or less cashless economy. This trend has been running a while, but it sped up greatly during the early phase of Covid when shopkeepers were reluctant to handle cash. We've since learned that the health risk to touching paper money was greatly overblown. Still, many of us quickly adjusted to carrying plastic only. And honestly, I have to admit that it's really convenient. Right up until the second that it isn't.

Credit card and ATM networks depend on connectivity. We've all had a waiter swipe our cards on their payment tablet, used a mobile-pay credit card reader attached to a vendor's phone, or just "tapped out" at the department store register. All those methods of accepting cards depend on the card reader being able to connect to the credit card's processing center to verify the card and process the charge. If the grid is down, none of those things can happen. But it doesn't require the grid shutting down to shut you out of credit card purchases. A power outage at the local grocery store would quickly put an end to credit card purchases. So would simply being in a dead zone.

I learned the last bit the hard way when I tried to rent a kayak at Pine Grove Furnace State Park. This beautiful park surround by Michaux state forest has several mountain lakes for recreation, nestled into the tip of the Blue Ridge Mountains. It also has no cell service in most of its near 700 acres. So I learned from the proprietor of the kayak rental shop. No cell service meant no credit cards accepted. Cash only. Naturally, I was at least a half hour drive from the nearest ATM. What should have been a relaxing day on the lake turned into a reminder to bring cash next time. I also make it a point now to check the websites of state and national parks along my route. You'd be surprised how many of our public parks are in dead zones. Often, that's a wonderful gift, providing a much appreciated digital detox. I sometimes select parks *because* they will force me to disconnect. I just like to know about it in advance.

Alternate communication devices

If your focus is off-grid living, you know without a doubt that you're not always going to have cellphone service. While that's part of the attraction of being off-grid, it can also become a safety issue in some circumstances. Depending on how much time you plan to spend off-grid, you may want to consider an alternative form of communication. You might also consider alternatives if you have a medical condition that could require emergency attention at some point.

When I travel, I have a satellite phone so I can get in touch with family if cell service isn't available. I keep it in my truck all the time, because you never know when you'll drive into a dead zone. Other communications devices you might consider are a CB radio, ham radio, shortwave radio, and a weather radio. I have a ham radio and a weather radio. The weather radio has rechargeable batteries, and it turns on to warn me when hazardous weather is approaching. These different types of technology take time to learn and money to acquire. Figure out what best fits your situation and pocketbook.

Like your trips, your communication is something you need to plan in advance.

Write It Down

Remember in Chapter 3 when we talked about route planning and the importance of having a paper map just in case you can't get to the Internet? Maps aren't the only thing you should bring along on paper. You should also have hardcopies with the addresses and phone numbers of every place you have reservations or plans to stay. Don't assume you'll be able to search for their numbers on your cell if you get lost.

Speaking of cellphones, when's the last time you memorized someone else's phone number? Be honest. It was probably a good decade or more. For safety's sake, you'll want to keep a paper copy of important phone numbers in both your RV and your towing vehicle. Include any numbers you would need in an emergency. I keep a phone list of towing companies, my car insurance agent, my closest friends, and the relatives I'm in constant contact with. Calling my brother might not be an actual emergency, but he's going to be worried beyond belief if he doesn't hear from me at least once a week.

Once you're written down those numbers? Make sure you use them, especially if you travel alone. Someone should know where you are, where you're going, and where you're supposed to be at any given time.

Ask Yourself

- Do I know where I'm going?
- Have I printed out the directions, addresses, and phone numbers of every place I've reserved?
- Do I have adequate supplies for basic power? Propane (if needed)? Extra gas? Batteries?
- Do I have a jump starter and a portable tire inflator?
- Am I set for backup power? Solar panels? A generator?

- Is my water tank full? My dark tank empty?
- Have I checked the lug nuts (just in case!)?
- Is my RV kitchen stocked with adequate food?
- Do I have emergency food in my towing vehicle?
- Did I double-check essentials in my towing vehicle? First aid kit? Water?
- Do I have safety equipment in the event of a breakdown? Flashlight? Flares? Reflective triangles?
- Am I prepared for cold weather in the event of a breakdown? Warm clothing? Blanket or sleeping bag?
- Can I communicate even without power? Have I stashed a charged power bank in my towing vehicle?
- Do I have enough cash to get by in a dead zone?
- Finally, am I safe from being locked out? Are there spare keys to my towing vehicle and RV?

AFTERWORD

I chose *Zen and the Art of Streaming* as a title because I believe having peace and joy in our lives is a choice we make each day. It doesn't matter if you are streaming down the road or camping somewhere, it's a choice.

I don't know about you, but when something goes wrong, I can feel negative energy creeping right up the back of my neck. When I broke down with my truck and Airstream, I felt a surge of fear and negative energy. Luckily, that surge was followed by a pause—and the absolute conviction that I refuse to be miserable no matter what. So instead of wallowing in the stress and anxiety of my experience, I chose to see that near miss as a priceless opportunity to look at exactly what I'd need to do to ensure that my future streaming adventures were as safe and peaceful as possible.

I'm not saying that was easy to do, mind you. Choosing joy takes practice. Most of all though, it takes planning. It's my hope that the planning laid out in this book will make it easier for you to choose joy by avoiding the most likely bumps on the Airstreaming road. I hope you enjoy the journey and I hope *Zen and the Art of Streaming* helps you set your mind and keep it set to choose joy!

If you have any questions along the way—please ask! In this book, I didn't include specific instructions for maintenance procedures. That information differs between styles and ages of Airstreams. It's also freely available. By choosing Airstream, you've also chosen a wealth of resources within the greater Airstream community. Use the forums. Explore the local clubs. They can put you in touch with both fellow Airstreamers and a wealth of information about RV living.

Also, don't be afraid to get your information directly from the horse's mouth. When I have questions about specific maintenance procedures, I go right to the Airstream site. They have just about everything I need there. If I have a unique question, I pick up the phone and call their technical support team. Those guys are amazing. They always answer my questions and are happy to help me. Let them help you too.

Enjoy the journey?

Time to get up and go!

See you in the Land of Abundance.

APPENDIX

RV Campground Memberships

Here are some great resources to check out.

1. **Airstream Club International**
   Airstream owners share their passion for the road and rally together. AirstreamClub.org

2. **Boondockers Welcome**
   This is a network of 2,000 plus property owners that allow RV owners to stay on their private property. Boondockerswelcome.com

3. **Escapees RV Club**
   Discounts at over 800 RV parks. Birds-of-a-feather groups with various interest like hiking, Christian, amateur radio, etc. Escapees.com

4. **Explorer RV Club**
   They offer discounts to parks in the United States and Canada. They also offer discounts on hotels, Emergency Roadside Service, RV dealers and campgrounds. Explorerrvclub.com

5. **Family Motorcoach Association (FMCA)**
   This membership helps connect RV owners with other individuals through specific events. They also have a roadside assistance service called FMC assist. Fmca.com

6. **Good Sam Club**
   They offer discounts at over 2,000 campgrounds. They also offer discounts at major camping retailers like Camping World, Gander Outdoors and Overton's. Goodsam.com

7. **Happy Camper Half Price Camping Club**
   You receive a 50% discount at hundreds of campgrounds in their network. Camphalfprice.com

8. **Harvest Hosts**
   This membership is comprised of a network of unique locations in North America that allow you to stay at their business in exchange for purchasing something at their location (3,422+ wineries, breweries, and farms). Harvesthosts.com

9. **KOA Rewards**
   They do a good job at keeping up their campgrounds and offer good amenities. You also earn rewards that can be used for free nights and 10% off the daily registration fee. Koa.com

10. **Overnight RV Parking**
    A subscription with over 14,000 parking locations. Just in case you get into a situation where you need to stop somewhere for the night without the camping fee. Overnightrvparking.com

11. **Thousand Trails**
    You pay a yearly fee to camp in a specific zones; Southwest, Northwest, Midwest, Northeast and Southeast. I got a free one-year membership when I purchased my Airstream from RVOne Airstream of Tampa. A great gift and an awesome membership. Thousandtrails.com

Made in the USA
Las Vegas, NV
21 January 2026

39923269R00046